Farm to Fork

How Your Produce
Impacts the Planet

Douglas B Sims, PhD

Douglas B Sims, PhD

Table of Contents

Acknowledgements

I am deeply grateful to my wife, whose unwavering support, wisdom, and love have been the foundation of my life for over 34 incredible years. You have been my steadfast anchor and greatest inspiration, not only throughout this process but in every step of our journey together.

To our two children, thank you for filling our lives with joy, teaching us the true essence of parenting, and being a constant source of pride and learning. Watching you grow has been one of life's most rewarding experiences.

To our extended family, your unwavering support and encouragement have been a cornerstone of our journey. Thank you for walking beside us and for being a vital part of this adventure.

I am also profoundly thankful to my friends and colleagues, whose experiences, feedback, and willingness to share insights have added depth and richness to this book. Observing and learning from your lives, as well as sharing my own, has been both enlightening and invaluable. Your contributions have shaped this work in countless ways, and I am truly appreciative of your generosity and openness.

Finally, I wish to acknowledge the many people I have encountered—whether in local malls, theaters, or across the world—whose everyday interactions and behaviors have provided profound inspiration. These moments of observation and reflection have enriched this book, offering real-world perspectives that made its content all the more meaningful.

Thank you all for being an integral part of this journey. Your presence, directly or indirectly, has made this work possible.

Forward

The food system sustains life itself, but its hidden costs are overwhelming our planet. The carbon cost of feeding a global population is staggering. Agriculture and food production account for roughly a quarter of global greenhouse gas emissions, driven by the sprawling, resource-intensive processes that define the modern food chain. From methane emissions from livestock to the fossil fuels required for transportation and refrigeration, the food we eat is deeply intertwined with the warming of our planet. The enormity of these challenges cannot be overstated.

This book lays bare the vast carbon footprint of our global food chain, unraveling its complexities and exposing the profound environmental impact of a system that prioritizes convenience, scale, and profit over sustainability. Every step of the food chain—from soil preparation and planting to processing, packaging, and shipping—carries an environmental cost. Imported avocados, out-of-season strawberries, and individually wrapped snacks are emblematic of a food system that spans continents but leaves behind an outsized carbon legacy. The distance food travels, the energy required to keep it fresh, and the waste generated along the way highlight the inefficiencies and unsustainable practices embedded in the system.

Yet, while this book confronts the enormity of the carbon burden, it also illuminates a path forward. It presents a hopeful, solution-driven vision for rethinking how we produce, distribute, and consume food. The solutions outlined here—precision agriculture, renewable energy on farms, reduced food waste, and a shift toward plant-based diets—are practical, achievable, and already making an impact in many parts of the world. Importantly, it underscores the collective power of consumers, producers, and policymakers to reimagine the food system as one that prioritizes environmental stewardship without sacrificing affordability or access.

This is a book for anyone who has ever wondered about the true cost of the food on their plate. It challenges us to confront the uncomfortable truths about the environmental toll of our global food chain but offers tangible ways to transform it into a system that works with the planet, not against it. It is both a wake-up call and a call to action—a reminder that the choices we make today will shape the legacy we leave for future generations.

As you read this book, you will be confronted with the scale of the problem, but you will also be empowered by the solutions. The future of our food chain, and the health of our planet, depend on our willingness to change, innovate, and act. Let this book be your guide to understanding the carbon cost of our food system and your inspiration to be part of the solution. Together, we can transform the way we feed the world while safeguarding the planet that sustains us all.

Chapter 1

The Modern Food Chain in Context

The modern food supply chain is an intricate, global system that connects producers, processors, distributors, retailers, and consumers in a vast network responsible for feeding billions of people worldwide (Smith & Williams, 2018). This chain operates on an unprecedented scale, stretching across continents and involving countless processes and technologies to ensure that food products reach consumers efficiently and safely. From farm to table, this complex web of activities encompasses a wide range of actors who play essential roles at every stage of the chain. Producers, such as farms, fisheries, and aquaculture operations, form the foundation of the supply chain, generating the raw materials needed for a vast array of food products. These producers operate under varying conditions, from small, family-owned farms to expansive, industrial-scale operations, each impacted by factors like climate, soil quality, and access to resources (Johnson & Brooks, 2020).

Once harvested, raw materials undergo transformation through food processing and manufacturing. Here, raw ingredients are converted into consumable products, a step that often involves cleaning, sorting, cutting, cooking, or preserving foods to extend shelf life, enhance flavor, and improve nutritional value (Lee & Thomas, 2019). Processing can range from simple treatments to complex processes that rely on advanced machinery and technology. For example, modern food processors use methods such as pasteurization, freeze-drying, and vacuum-packing to maintain freshness and quality over long distances. These steps are essential in meeting global consumer demands for convenience and variety, as processed foods are easier to distribute, store, and prepare.

The distribution and logistics stage ensures that food products, now ready for sale, are transported from processing plants to various retail outlets worldwide. This movement involves a sophisticated logistical network that includes refrigerated transport, warehousing, and cross-border coordination to maintain quality and prevent spoilage (Brown, 2017). The global food supply chain's reliance on transportation means that it's highly vulnerable to disruptions. Fuel price fluctuations, labor shortages, or environmental crises can delay deliveries, affect product availability, and increase costs. Moreover, the carbon footprint associated with global food transportation is significant, raising concerns about the environmental impact of extensive supply chains (Miller, 2021).

Retailers, ranging from small neighborhood stores to large multinational supermarket chains, represent the final link before food reaches consumers. These retailers act as intermediaries, bringing food products to a wide variety of consumers with differing needs, budgets, and dietary preferences. Consumer behavior greatly influences the retail sector, pushing food suppliers to adapt to preferences for local, organic, or sustainably sourced products. With the rise of online shopping and delivery services, many retailers now utilize digital platforms to connect with customers, expanding the chain's reach even further.

Ultimately, consumer preferences and demands drive much of the system's behavior, determining which foods are grown, how they are packaged, and how quickly they must be delivered to meet demand. Consumer trends, such as the increasing desire for fresh, organic, and sustainable products, shape agricultural practices and influence retail strategies (Miller, 2021). The demand for convenience has also fostered growth in pre-packaged and ready-to-eat foods, altering food manufacturing and packaging processes. This consumer-driven system is highly dynamic, with suppliers and retailers constantly adapting to shifts in dietary trends, health awareness, and ethical concerns.

Despite its efficiency, the global food chain faces significant challenges, including vulnerabilities to global disruptions such as pandemics, political conflicts, and natural disasters, which can lead to shortages and increased prices (Roberts & Klein, 2019). For example, the COVID-19 pandemic exposed significant weaknesses in the supply chain, from labor shortages on farms to bottlenecks in shipping and distribution, leading to widespread food insecurity and increased consumer prices. Additionally, food safety and quality standards are crucial in maintaining public health, requiring rigorous monitoring and advanced technologies, such as cold storage and supply tracking, to ensure that food remains safe throughout the journey from production to consumption (Adams, 2015). These technologies are essential in preventing contamination and spoilage, particularly in perishable items like meat, dairy, and fresh produce.

In summary, the modern food supply chain is a highly efficient yet complex system, shaped by the demands of a globalized economy and driven by advancements in agricultural and logistical technology. However, it is also vulnerable to a range of disruptions and requires constant innovation and oversight to ensure that food safety, quality, and accessibility are maintained across all stages.

Historical Perspective: How the Food Chain Evolved with Industrialization

The evolution of the global food chain can be traced back to the Industrial Revolution, a period that fundamentally transformed food production, distribution, and consumption on a massive scale (Thompson, 2018). Before the onset of industrialization, most communities relied on local, subsistence-based food systems, which involved small-scale agriculture, local trades, and a reliance on seasonal cycles to meet food needs (Green, 2016). This localized approach meant that food was primarily consumed where it was produced, with minimal reliance on external sources. Regional food trade was limited by transportation constraints, as fresh food could not travel far without

3

spoiling. Community members relied on traditional farming methods and cooperative food sharing, emphasizing sustainability and resource management based on the limitations of their immediate environment.

The Industrial Revolution, which began in the 18th century, introduced groundbreaking technologies and innovations that reshaped agriculture, enabling an unprecedented leap in food production capabilities (Garcia, 2017). Mechanized farming equipment such as the steel plow, seed drill, and steam-powered tractors allowed farmers to cultivate larger areas of land with greater efficiency. These advancements reduced the reliance on human and animal labor, paving the way for large-scale agriculture. Chemical fertilizers and pesticides were also introduced, further boosting crop yields and minimizing losses from pests and disease. These changes made it possible to grow a larger quantity and variety of crops, effectively meeting the demands of a rapidly growing urban population, where industrial workers were concentrated and dependent on reliable food sources.

Another significant development during this era was the improvement of transportation infrastructure, which transformed the way food was distributed. The advent of the railroad and later the use of steamships allowed food to be transported over long distances at unprecedented speeds, expanding the reach of food producers to new markets and consumers across regions and countries (Lopez, 2019). This improvement in transportation led to the development of a broader food market and, for the first time, made it possible to supply urban areas with a steady flow of food from distant agricultural regions. Perishable items like meat and dairy could now be shipped across the country with the help of refrigerated train cars, which were introduced in the late 19th century, extending food shelf life and opening up new trade routes.

The mid-20th century ushered in the Green Revolution, a period characterized by rapid agricultural advancements aimed at addressing global food shortages and supporting population growth, especially in developing countries. New high-yield crop varieties were developed, and the use of synthetic fertilizers and pesticides became widespread, leading to an increase in food production per acre (Anderson, 2015). Irrigation systems were expanded, and modern agricultural techniques, such as monocropping and mechanized harvesting, became the standard. While the Green Revolution was largely successful in preventing famines and increasing the food supply, it also had unintended consequences. The heavy reliance on chemical inputs led to soil degradation and water pollution, while intensive farming practices contributed to biodiversity loss. Additionally, many small-scale farmers struggled to keep up with the costs of new technologies, which favored large-scale agribusinesses, creating a growing divide between corporate and family-owned farms.

By the late 20th century, the food supply chain had become increasingly dominated by multinational corporations, which began to exert significant control over food production, processing, and distribution. Companies like Nestlé, Monsanto, and Cargill expanded their influence, not only through agricultural production but also by developing proprietary seeds, pesticides, and fertilizers, which became essential for high-yield farming (Johnson & Brooks, 2020). These corporations leveraged economies of scale, centralized operations, and global distribution networks, enabling them to provide vast quantities of food products to consumers worldwide. However, the concentration of power in a few large corporations also raised concerns over the impact on small farmers, who found it increasingly difficult to compete with corporate-scale operations. Moreover, this concentration of control often led to homogenized food products and reduced biodiversity, as corporations focused on high-yield, profitable crops rather than preserving a diversity of plant and animal species.

Today's food chain, shaped by these historical forces, represents the legacy of industrialization—a system that is highly efficient but fraught with environmental, economic, and social implications (Smith & Williams, 2018). Industrialized food production has increased food availability and reduced costs for many consumers, but it also relies heavily on nonrenewable resources, such as fossil fuels for transportation and synthetic fertilizers derived from petrochemicals. This dependence on industrial inputs contributes to environmental issues, including greenhouse gas emissions, soil depletion, and water pollution, raising questions about the long-term sustainability of the food supply. Additionally, the global nature of today's food chain makes it vulnerable to disruptions, as seen during the COVID-19 pandemic, when labor shortages, transportation delays, and export restrictions caused widespread food insecurity and price increases (Roberts & Klein, 2019). The modern food system, while effective at meeting global demand, requires constant adaptation to balance efficiency with environmental stewardship and social equity, highlighting the need for sustainable practices to address the challenges posed by this industrialized legacy.

Defining Costs: Economic, Environmental, and Social

The costs of the modern food chain extend far beyond the economic transactions involved in bringing food to market, impacting not only global economies but also the environment, social equity, and public health. Economically, the global food chain affects food prices and affordability worldwide, but particularly in low-income regions where import costs and market demands can inflate prices beyond the reach of local consumers. For communities that rely on imported food, fluctuations in global markets can severely impact food affordability, leading to increased vulnerability and food insecurity (Brown, 2017). In regions where food is produced for export, local populations may face food shortages and inflated prices as resources are directed toward high-profit foreign markets rather than domestic needs. Furthermore, employment within the food sector remains a critical component of many economies, employing millions worldwide. However, wage disparities and labor exploitation are prevalent, especially among agricultural workers and food processing laborers, who often work in demanding, low-paying conditions with limited job security (Lee & Thomas, 2019). The profit-driven nature of industrialized food production often pressures employers to minimize labor costs, leading to systemic issues like low wages, inadequate benefits, and poor working conditions, which contribute to poverty and inequality within the sector.

Environmentally, the food chain exacts a substantial toll on natural resources, with industrial farming practices and global transportation systems contributing to a range of ecological problems. Land use changes and deforestation to accommodate large-scale agriculture are major contributors to biodiversity loss, as natural habitats are cleared for monoculture crops and livestock production (Miller, 2021). This deforestation disrupts ecosystems, reducing wildlife populations and threatening species with extinction. Water resources are also heavily impacted, as modern agriculture is one of the largest consumers of freshwater globally, leading to water scarcity in many regions. The intensive use of chemical fertilizers and pesticides in industrial farming contaminates soil and water sources, resulting in pollution that affects ecosystems and human health. Additionally, the global distribution network, reliant on fossil fuels, generates significant carbon emissions, contributing to climate change. Industrialized farming practices, including mechanized equipment and synthetic inputs, further exacerbate greenhouse gas emissions, undermining efforts to reduce the carbon footprint of food production (Garcia, 2017). Soil degradation is another consequence, as intensive farming depletes soil nutrients, reduces organic matter, and increases erosion, ultimately lowering the land's productivity and long-term agricultural potential.

Socially, the modern food chain raises a multitude of concerns related to food security, public health, and cultural heritage. Food security remains a pressing issue, as global inequalities in food distribution mean that some communities face chronic shortages while others have an overabundance. In many parts of the world, limited access to fresh, nutritious food results in malnutrition and a reliance on imported, processed foods that lack essential nutrients. This disparity is starkly visible in urban food deserts, where low-income neighborhoods have limited access to affordable, healthy foods, forcing residents to depend on fast food and convenience store items (Lopez, 2019). Meanwhile, the prevalence of highly processed, calorie-dense foods in developed countries has led to rising rates of obesity, diabetes, and other diet-related illnesses, creating a public health crisis. The modern food chain, with its emphasis on efficiency and uniformity, has also diminished local food cultures and traditional agricultural practices. As global food corporations expand, traditional and diverse food practices are increasingly replaced by standardized products, leading to the erosion of cultural identities and culinary heritage (Green, 2016). The shift toward homogenized diets and mass-produced foods not only impacts cultural diversity but also reduces agricultural resilience, as local varieties are replaced by a limited range of commercially viable crops.

These cumulative economic, environmental, and social costs underscore the urgent need for sustainable solutions that address the complex interplay of forces shaping the global food chain today. Sustainable agricultural practices, such as regenerative farming, agroecology, and crop diversification, can help mitigate environmental degradation by improving soil health, enhancing biodiversity, and reducing reliance on chemical inputs. Policy interventions, such as fair labor practices, living wages, and support for small-scale farmers, are essential to addressing social inequities within the food sector. Furthermore, promoting local food systems, reducing food waste, and encouraging responsible consumer choices can collectively alleviate some of the economic and environmental burdens of the global food chain. Ultimately, reshaping the food chain to prioritize sustainability, equity, and resilience will require a collaborative approach among governments, corporations, communities, and individuals. By addressing these interconnected challenges, the modern food system can evolve to meet the needs of a growing global population without compromising ecological balance, social well-being, or cultural diversity (Anderson, 2015).

Chapter 2

The Foundation of the Agriculture and Carbon Footprints

Industrial agriculture, which focuses on maximizing output through large-scale monocropping, mechanization, and the use of synthetic inputs, has become the dominant approach in modern food production. This method prioritizes efficiency and yield above all else, catering to a growing global population and the high demands of food markets, especially in urban areas. Large-scale monocropping allows for streamlined production and harvests of single crops like corn, soy, and wheat, but it sacrifices crop diversity, making the agricultural landscape more vulnerable to pests, diseases, and environmental changes. Additionally, mechanization enables vast tracts of land to be farmed with fewer labor inputs, lowering production costs and making food more accessible to consumers. However, these advantages often come at a steep cost to environmental and social sustainability.

Industrial farming relies heavily on chemical fertilizers, pesticides, and intensive irrigation, which, while boosting yields, pose significant risks to soil health, water quality, and local ecosystems. Chemical fertilizers, primarily nitrogen-based, are used to enrich the soil with nutrients that monocultures exhaust over time. However, excess nitrogen and other compounds often leach into nearby water sources, contributing to nutrient pollution, which can lead to harmful algal blooms and "dead zones" in aquatic ecosystems (Garcia, 2018). Pesticides, designed to control pests and diseases, can also harm non-target species, including beneficial insects, birds, and aquatic life, creating a ripple effect in the ecosystem. Intensive irrigation required for crops in arid or semi-arid regions depletes freshwater resources, strains aquifers, and, in some cases, leads to soil salinization, further reducing land fertility over time. Furthermore, industrial agriculture contributes significantly to greenhouse gas emissions, both from fossil fuel use in mechanized farming equipment and the release of nitrous oxide from fertilizers, a potent greenhouse gas (Garcia, 2018).

Another major drawback of industrial farming practices is the promotion of monocultures. By focusing on single-crop production, industrial agriculture reduces biodiversity and creates a highly uniform agricultural landscape. This lack of diversity leaves crops more susceptible to diseases and pest outbreaks, as there are fewer natural barriers or pest predators within monoculture fields. For instance, when a pest or disease strikes a single-crop farm, the entire field is at risk, potentially leading to widespread crop failure and economic loss. Additionally, monocultures diminish habitat diversity, reducing the resilience of surrounding ecosystems and driving declines in pollinator populations, which are essential for many crops (Brown, 2019). These practices have raised concerns over the long-term viability of industrial agriculture, as they lead to an over-reliance on chemical inputs and increasingly unstable ecosystems.

In contrast, sustainable farming practices seek to produce food in a manner that is both environmentally sound and socially responsible. Rather than prioritizing short-term yield gains, sustainable methods focus on maintaining soil health, fostering biodiversity, and reducing the need for synthetic inputs. Practices like crop rotation and polycultures increase biodiversity within farming systems, breaking pest and disease cycles naturally and reducing the need for chemical pesticides. For instance, crop rotation involves alternating different crops in the same field across seasons, which disrupts the habitat of pests and pathogens that target specific crops (Lee & Carter, 2020). By planting complementary species that can repel pests or attract beneficial insects, polycultures help control pest populations without harming the broader ecosystem.

Another key element of sustainable agriculture is regenerative farming, a holistic approach that emphasizes restoring and enhancing soil health. Regenerative practices, such as composting, cover cropping, and reduced tillage, contribute to building soil organic matter, which improves soil fertility, water retention, and carbon sequestration. Composting recycles organic waste into nutrient-rich soil amendments, reducing the need for synthetic fertilizers and enhancing soil structure. Cover crops, planted during off-seasons, prevent soil erosion, suppress weeds, and improve soil nitrogen levels through natural processes (Anderson, 2019). Reduced tillage, or no-till farming, minimizes soil disturbance, which preserves soil microorganisms and organic matter, ultimately enhancing soil's ability to store carbon. These practices not only support healthier crops but also play a role in mitigating climate change by sequestering carbon in the soil.

Sustainable farming also considers the social aspects of agriculture, striving to support local economies and protect farm workers. Unlike industrial agriculture, which often consolidates profits within large agribusiness corporations, sustainable farming models—such as community-supported agriculture (CSA) and small-scale organic farms—encourage local food production and distribution, which strengthens local economies. Sustainable practices also emphasize fair labor conditions and prioritize the health and safety of farm workers by minimizing exposure to harmful chemicals. By fostering strong connections between farmers and their communities, sustainable agriculture promotes food security, transparency, and resilience in the face of global market fluctuations.

In the context of climate change, sustainable agriculture is increasingly recognized as a viable alternative to industrial farming, as it offers methods to reduce agriculture's carbon footprint and enhance resilience to climate variability. Climate-resilient practices, like agroecology and agroforestry, integrate trees and diverse crops within agricultural landscapes, creating microclimates that protect crops from extreme weather and increase biodiversity. Agroecology combines ecological principles with farming to create a diverse, self-sustaining system that minimizes external inputs and adapts to local conditions. Agroforestry integrates trees into crop and livestock systems, offering benefits such as soil stabilization, shade, and carbon storage. These approaches not only reduce environmental impact but also promote food security by fostering diversified production systems that are less vulnerable to climate change and market shocks (Lee & Carter, 2020).

Overall, sustainable agriculture represents a path forward, balancing the need for productivity with the imperative to protect ecosystems and communities. While industrial agriculture has provided the world with a reliable food supply, its environmental and social costs are increasingly apparent. Transitioning to sustainable farming practices will require a paradigm shift in agricultural policy, consumer preferences, and economic incentives, but it offers the potential for a more resilient and equitable food system that supports both people and the planet in the long term.

The Carbon Impact of Crop and Livestock Farming

Agriculture is a major contributor to global greenhouse gas emissions, with both crop and livestock farming responsible for significant carbon footprints that impact climate change on a global scale. Crop farming emits greenhouse gases through a variety of processes, the most notable being land-use changes, fertilizer application, and fuel consumption in agricultural machinery. One of the most substantial sources of emissions within crop farming is land conversion, where forests, grasslands, and other ecosystems are cleared to create farmland. This conversion, particularly deforestation for expanding agricultural lands, releases large amounts of carbon that was previously stored in vegetation and soil, making it one of the leading sources of carbon emissions in the agricultural sector (Roberts, 2017). When forests are cleared, carbon that has been sequestered in trees, plants, and soils for decades or even centuries is released back into the atmosphere as CO_2. Additionally, these cleared lands, often used for monoculture crops, disrupt the local carbon cycle, reducing the land's capacity to store carbon and further exacerbating climate change.

Fertilizer application is another significant contributor to greenhouse gas emissions in crop farming. Fertilizers, especially nitrogen-based ones, are essential for enhancing crop yields; however, they release nitrous oxide (N_2O), a potent greenhouse gas with a global warming potential approximately 300 times greater than CO_2 (Miller, 2019). When nitrogen fertilizers are applied to crops, some of the nitrogen converts to N_2O, which then escapes into the atmosphere, contributing to climate change. Moreover, the production of synthetic fertilizers is itself an energy-intensive process that relies on fossil fuels, further adding to the sector's emissions footprint. The use of agricultural machinery, which relies heavily on fuel, contributes additional CO_2 emissions, as tractors, harvesters, and irrigation systems consume vast amounts of energy for planting, maintaining, and harvesting crops. These emissions collectively contribute to the substantial carbon footprint associated with crop farming.

Livestock farming, particularly cattle production, has an even more intense impact on greenhouse gas emissions due to the production of methane (CH_4), a greenhouse gas with a global warming potential significantly higher than CO_2. Methane is released through enteric fermentation in ruminant animals like cows, sheep, and goats, where microbes in their digestive systems break down food and produce methane as a byproduct (Smith & Lopez, 2021). This methane is then expelled primarily through belching, accounting for a substantial portion of emissions from animal agriculture. In addition to enteric fermentation, methane is also produced from manure management, as livestock waste decomposes under anaerobic conditions in manure lagoons and storage pits. When not properly managed, this manure emits both methane and nitrous oxide, further contributing to livestock farming's environmental impact.

The environmental impact of livestock farming extends beyond direct emissions. Raising livestock requires significant resources, including land, water, and feed, all of which contribute indirectly to greenhouse gas emissions and environmental degradation. For instance, the production of feed crops, such as corn and soy, demands substantial amounts of land, often leading to deforestation and land-use change that releases stored carbon and reduces biodiversity (Johnson, 2020). Additionally, the water footprint of livestock farming is considerable, as animals require large quantities of water for drinking, and water is also needed to grow their feed. This water usage puts additional pressure on freshwater resources, especially in regions prone to droughts and water scarcity.

The deforestation associated with livestock farming, both for grazing and feed production, has far-reaching ecological implications. Forests, especially tropical rainforests, are among the most effective carbon sinks on the planet, sequestering large amounts of CO_2 from the atmosphere. When these forests are cleared for agriculture, not only is this stored carbon released, but the land also loses its ability to sequester carbon in the future. Deforestation for livestock agriculture contributes not only to increased carbon emissions but also to significant biodiversity loss, as forested areas are home to countless plant and animal species. The destruction of these habitats threatens biodiversity, disrupts ecosystems, and reduces the resilience of these areas to environmental changes.

Given the substantial environmental impacts associated with traditional agricultural practices, there is a growing recognition of the need to reduce the carbon footprint of agriculture, particularly through sustainable livestock management and alternative protein sources. Sustainable livestock management includes practices such as rotational grazing, improved manure management, and optimizing feed to reduce methane emissions from enteric fermentation. Rotational grazing involves moving livestock between different pastures to allow vegetation to recover, which can improve soil health, reduce erosion, and increase the land's carbon sequestration potential. Improved manure management, such as composting or using anaerobic digesters, can capture methane emissions and even convert them into renewable energy, reducing the overall emissions from livestock farms.

Furthermore, alternative protein sources, such as plant-based proteins, lab-grown meat, and insect protein, are being explored as viable solutions for meeting protein demands while reducing environmental impacts. Plant-based proteins, derived from sources like soy, peas, and lentils, require fewer resources and generate far fewer emissions than animal-based proteins. Lab-grown meat, though still in its early stages of development, has the potential to deliver the taste and nutritional benefits of conventional meat with a significantly lower environmental footprint. Insect protein, already popular in many cultures, provides a highly efficient source of protein that requires minimal land, water, and feed compared to traditional livestock.

Reducing agriculture's carbon footprint will require a multifaceted approach that includes adopting sustainable practices, improving resource efficiency, and rethinking protein production to meet the demands of a growing population. Sustainable farming practices offer a pathway to mitigate the environmental impacts of agriculture, while alternative protein sources provide an opportunity to shift dietary habits in a way that lessens the pressure on natural resources. As the world grapples with climate change, finding ways to make agriculture more sustainable is essential for ensuring both environmental health and food security in the future (Garcia, 2018).

Costs Associated with Soil Degradation, Deforestation, and Biodiversity Loss

The environmental costs of industrial agriculture are vast, with soil degradation, deforestation, and biodiversity loss standing out as some of the most critical issues. Soil degradation results from industrial practices that prioritize short-term yield over long-term soil health, including over-tilling, monocropping, and the excessive use of chemical fertilizers. These practices disrupt the natural composition of soils, stripping them of essential nutrients and organic matter and ultimately reducing their fertility and productivity over time (Brown, 2019). Over-tilling, which disturbs soil structure, accelerates erosion and depletes soil organic matter, essential for retaining moisture and nutrients. Monocropping, the repeated cultivation of the same crop over large areas, not only exhausts specific nutrients but also weakens soil resilience, as it lacks the diversity that naturally replenishes and strengthens soils. The heavy reliance on chemical fertilizers in industrial agriculture exacerbates this problem; while fertilizers temporarily boost yields, they often lead to nutrient imbalances in the soil. Over time, these imbalances can harm soil microbiota, the organisms that contribute to natural soil fertility, making the soil increasingly dependent on synthetic inputs. Consequently, industrial agriculture often faces diminishing returns, with farmers needing to apply even more fertilizers and pesticides to maintain yields as soils degrade.

Intensive farming practices also lead to other forms of soil degradation, including soil erosion, salinization, and nutrient depletion. Soil erosion occurs when the protective vegetative cover is removed, leaving topsoil vulnerable to wind and water runoff. This topsoil, rich in nutrients and organic matter, is essential for plant growth, and its loss has devastating effects on agricultural productivity. Erosion reduces crop yields and decreases the land's potential to store carbon, reducing its ability to act as a carbon sink (Lee & Carter, 2020). Salinization, another consequence of industrial irrigation practices, occurs when irrigation water evaporates, leaving behind salts that accumulate in the soil. Over time, high salinity levels can render soil infertile, making it unsuitable for crop cultivation. The cumulative effect of these forms of degradation is a loss of productive land, increased costs for farmers, and reduced resilience to climate change impacts, as degraded soils store less carbon and are more susceptible to drought and flooding.

Deforestation is another significant environmental cost associated with agricultural expansion, particularly in tropical regions where forests are cleared to make way for cash crops like palm oil, soy, and cocoa, as well as for livestock grazing. This large-scale deforestation is driven by the global demand for these commodities, which pushes farmers and corporations to convert forested areas into farmland. The loss of forests has far-reaching ecological impacts, as forests are not only vital carbon sinks, absorbing significant amounts of CO_2 from the atmosphere, but also serve as habitats for countless species (Roberts, 2017). When forests are cleared, the stored carbon in trees and soil is released back into the atmosphere, contributing to greenhouse gas emissions and accelerating climate change. Moreover, the removal of forested land disrupts local ecosystems, leading to a loss of biodiversity as countless species lose their habitats. This habitat destruction not only affects wildlife populations but also destabilizes ecosystems that provide essential services, such as pollination, water filtration, and climate regulation. The destruction of forests for agricultural expansion thus undermines both environmental stability and biodiversity, with consequences that are felt far beyond the immediate areas of deforestation.

Biodiversity loss, closely linked to both soil degradation and deforestation, is one of the most pressing environmental challenges associated with industrial agriculture. In pursuing high yields, industrial agriculture often relies on monocultures, the cultivation of a single crop over large areas, which reduces habitat diversity and disrupts natural ecosystems. Monocultures make agricultural systems more vulnerable to pests and diseases, as the lack of plant diversity means there are no natural barriers or predators to control pest populations. This reliance on monocultures and the intensive use of land for single crops weaken ecosystems and reduce the resilience of agricultural systems to environmental stresses, such as climate change and pest outbreaks (Anderson, 2019). Furthermore, industrial farming practices such as the extensive use of chemical pesticides and herbicides can harm non-target species, including pollinators like bees and butterflies, which play a crucial role in crop pollination. The decline of pollinators is a significant concern for global food security, as many crops rely on these species for reproduction.

The cumulative effects of biodiversity loss in agricultural landscapes also extend to soil health, as a reduction in plant and microbial diversity limits the natural processes that contribute to soil fertility. Diverse ecosystems support a wide range of organisms that contribute to nutrient cycling, pest control, and soil structure. When biodiversity is reduced, these ecosystem services are compromised, making agricultural systems increasingly dependent on external inputs like fertilizers and pesticides. This dependence on synthetic inputs not only raises production costs for farmers but also increases the environmental footprint of agriculture, as these chemicals often leach into waterways, causing pollution and harming aquatic ecosystems.

In response to these environmental challenges, sustainable farming approaches that incorporate biodiversity offer promising solutions for protecting ecosystems while maintaining productivity. Agroforestry, a practice that integrates trees and shrubs into crop and livestock systems, enhances biodiversity by creating diverse habitats that support various species and improve soil health. Trees in agroforestry systems contribute to carbon sequestration, protect soil from erosion, and enhance water retention, making agricultural landscapes more resilient to environmental stresses. Polycultures, the cultivation of multiple crop species in the same area, promote ecological diversity and reduce pest and disease pressures by creating complex habitats that discourage the spread of pests. These biodiversity-focused farming practices help to restore ecological balance, reduce the need for chemical inputs, and improve the sustainability of agricultural systems.

The environmental costs of industrial agriculture—soil degradation, deforestation, and biodiversity loss—highlight the urgent need to adopt sustainable agricultural practices that prioritize ecological health and resilience. Sustainable practices such as agroforestry and polycultures provide a pathway toward a more balanced agricultural system that values ecosystem services and long-term productivity. As the global demand for food continues to rise, transitioning to these sustainable practices is essential for protecting natural resources, mitigating climate change, and ensuring the resilience of food systems for future generations.

Chapter 3

Water Usage and Food Production Costs

Agriculture is the world's largest consumer of freshwater, accounting for approximately 70% of global water withdrawals (Smith, 2019). This high demand for water in agriculture highlights the sector's "water footprint," which encompasses the volume of water used throughout every stage of the production process, from initial crop irrigation to livestock watering. The water footprint of agriculture is extensive and multifaceted, with significant water resources consumed directly in crop fields and indirectly through feed production for livestock. In regions with limited rainfall, such as arid and semi-arid zones, irrigation is indispensable for crop growth. Without it, agricultural productivity would be severely limited, affecting both local food security and global food supplies. However, many traditional irrigation methods are notoriously inefficient. For instance, flood irrigation, one of the oldest and most common techniques, involves flooding fields with water, much of which is lost to evaporation or runoff before it even reaches the plants' root systems (Garcia & Li, 2020).

To address these inefficiencies, modern irrigation techniques have been developed to maximize water use efficiency. Drip and sprinkler irrigation systems, for example, offer precise water delivery by targeting the root zones of plants, minimizing water wastage due to evaporation. Drip irrigation, which involves placing a network of tubes with small holes near the base of plants, delivers water slowly and directly to the roots. This method can save up to 60% more water compared to traditional flood irrigation, significantly reducing water usage in agriculture (Anderson, 2018). Sprinkler irrigation, while less efficient than drip irrigation, still offers improvements over flood irrigation by dispersing water over crops in controlled amounts. These systems help mitigate water loss and make irrigation feasible in water-scarce regions where efficient water usage is essential for crop survival. Moreover, advanced technologies, such as soil moisture sensors and automated irrigation systems, allow farmers to monitor and adjust water levels based on real-time conditions, further enhancing water use efficiency.

Beyond crop irrigation, another critical aspect of agriculture's water footprint is water wastage throughout the supply chain. Water wastage occurs at multiple points, including processing, packaging, and transportation, adding to agriculture's already substantial water demands. Food processing facilities, for instance, use large quantities of water to wash, sort, and package food products, and much of this water is discarded after use, often without proper treatment, which exacerbates wastage (Lee, 2019). These facilities process everything from fruits and vegetables to grains and dairy, each requiring specific cleaning and preparation steps that consume significant water volumes. The waste generated in these processes is not always recycled or reused, contributing to a higher water footprint for processed foods. Additionally, transportation and storage of agricultural products involve further water use, particularly for cooling and preserving perishable goods, which require regular misting or refrigeration.

Livestock farming also has a considerable water footprint, as water is essential not only for the animals themselves but also for the production of animal feed. Animals need water for drinking, sanitation, and temperature regulation, especially in hotter climates where cooling is necessary to prevent heat stress. Furthermore, the production of feed crops, such as corn and soybeans, requires substantial irrigation, particularly in areas with limited natural rainfall. This indirect water use significantly contributes to the overall water footprint of meat production. For example, producing one kilogram of beef can require approximately 15,000 liters of water, largely due to the water-intensive nature of growing feed crops, compared to around 1,500 liters for one kilogram of wheat (Miller, 2020). The cumulative water requirements for livestock farming are therefore much higher than for plant-based foods, underscoring the environmental costs associated with meat production.

To address the extensive water demands of agriculture, a range of water conservation strategies have been developed, with the goal of reducing the sector's water footprint and preserving freshwater resources. Improving irrigation efficiency is a primary focus, with farmers increasingly adopting water-saving technologies and sustainable water management practices. Precision agriculture, for example, uses data from soil sensors, satellite imagery, and weather forecasts to determine optimal irrigation schedules, ensuring that crops receive the exact amount of water needed at the right times (Smith, 2019). This tailored approach not only conserves water but also reduces the costs associated with over-irrigation and runoff management. Technologies such as moisture sensors allow farmers to monitor soil hydration levels and adjust irrigation accordingly, minimizing water usage without compromising crop yields.

Additionally, conservation practices like rainwater harvesting and the reuse of treated wastewater are becoming more common, particularly in regions facing severe water scarcity. Rainwater harvesting involves capturing and storing rainwater for agricultural use, reducing the reliance on groundwater and surface water sources. In drought-prone areas, stored rainwater can provide a critical buffer during dry seasons, enabling farmers to continue irrigating crops when other water sources are limited. The reuse of treated wastewater for irrigation is another effective strategy, as it repurposes water that would otherwise be discarded. Treated wastewater can be safely used for certain crops, reducing the strain on freshwater supplies. Some countries, such as Israel, have implemented large-scale wastewater recycling programs for agriculture, demonstrating the potential of these practices to support sustainable water use.

Governments and international organizations are actively promoting water conservation practices in agriculture through subsidies, awareness programs, and regulations. For example, government programs may offer financial incentives to farmers who adopt water-efficient technologies, making it easier for small-scale farmers to transition to sustainable practices. Educational campaigns also play a vital role, raising awareness about the importance of water conservation and providing farmers with the knowledge needed to implement effective strategies. In regions experiencing chronic water shortages, regulations may limit the amount of water allocated for irrigation, encouraging farmers to adopt more efficient practices. Together, these efforts aim to promote sustainable water use in agriculture, helping to safeguard freshwater resources for future generations.

In summary, the water footprint of agriculture is a significant component of the sector's environmental impact, with water usage spanning irrigation, processing, and livestock production. While traditional irrigation methods have contributed to high levels of water wastage, modern techniques like drip and sprinkler irrigation, alongside conservation practices, offer promising solutions for reducing agriculture's water demands. The cumulative effects of these water-saving measures, supported by government initiatives and sustainable water management practices, have the potential to create a more resilient agricultural sector capable of meeting global food demands without depleting vital water resources. As climate change and population growth continue to increase pressures on water availability, adopting these water conservation strategies is essential for ensuring sustainable food production and preserving freshwater ecosystems.

How Water Shortages Affect Food Costs and Carbon Emissions

Water scarcity has a direct impact on food costs, as limited water resources lead to reduced agricultural productivity, increased input costs, and ultimately higher prices for consumers. When water is scarce, farmers are often forced to reduce irrigation, resulting in lower crop yields and, in severe cases, crop failures (Garcia & Li, 2020). These reductions in yield create supply shortages, which drive up prices for essential foods like wheat, rice, and vegetables. For instance, during prolonged droughts, such as the 2012 drought in the United States, crop yields dropped dramatically, causing a significant spike in food prices both domestically and globally (Anderson, 2018). Such price volatility can be particularly detrimental to food security in low-income regions, where households allocate a substantial portion of their income to food. For these communities, even small increases in food prices can limit access to basic nutrition, exacerbating poverty and malnutrition.

Water scarcity also has indirect impacts on agriculture's carbon footprint, contributing to climate change. When water becomes limited, farmers may rely on energy-intensive methods, such as groundwater extraction, to meet irrigation demands. Groundwater extraction, especially from deep aquifers, requires substantial energy inputs, further increasing agriculture's carbon emissions (Lee, 2019). In addition, the scarcity of water can lead to shifts in agricultural production to areas with more water resources, often requiring products to travel longer distances to reach markets, which increases transportation-related emissions. These shifts in production can disrupt established agricultural regions, increasing the overall environmental footprint of food production and underscoring the need for sustainable water management practices in agriculture.

The economic implications of water scarcity extend beyond the agricultural sector, affecting entire economies. Reduced agricultural output can lead to higher inflation rates, as food prices drive up overall costs of living. For countries that rely on agricultural exports, water shortages can decrease export revenues and create a need to import more food, leading to increased trade deficits. The ripple effects of water scarcity are therefore far-reaching, particularly in economies heavily dependent on agriculture for income and employment (Miller, 2020). In these contexts, water scarcity can undermine economic stability, reduce growth opportunities, and exacerbate poverty by limiting jobs in agricultural regions. Economic constraints, combined with rising food prices, often create a cycle where water scarcity deepens existing economic challenges and threatens the well-being of millions of people.

In response to the extensive impacts of water scarcity, governments and organizations are increasingly prioritizing sustainable water management practices. Measures such as regulated water usage, drought-resistant crop varieties, and improved irrigation technology are being implemented to help mitigate the effects of water scarcity on food production. These strategies are essential not only for maintaining agricultural productivity but also for reducing agriculture's carbon footprint and ensuring economic stability in water-scarce regions. Sustainable water management practices help secure food supplies, stabilize economies, and lower emissions, addressing the interconnected challenges posed by water scarcity in modern agriculture.

Case Studies of Drought-Prone Areas and Their Economic Impacts

The effects of water scarcity on agriculture and food production are particularly evident in drought-prone areas, where limited water availability has profound economic impacts. One notable example is California, a region known for producing high-value crops such as almonds, grapes, and citrus. California has experienced severe droughts in recent years, with water shortages disrupting agricultural production and leading to significant economic losses. The 2012-2016 drought, for instance, was particularly devastating for California's agricultural sector, which suffered an estimated $3.8 billion in direct costs, including losses from reduced crop revenue, increased pumping costs for groundwater, and fallowing of farmland to conserve resources (Garcia & Li, 2020). This drought affected not only farmers but also consumers, as prices for water-intensive crops such as almonds and grapes rose substantially, underscoring the interconnectedness of water scarcity, food production, and market prices.

Australia provides another critical example of the economic toll of drought on agriculture. The country has faced recurring droughts that have reshaped its agricultural practices and policies. The Millennium Drought, spanning from 1997 to 2009, led to drastic reductions in water allocations for farmers, particularly in the Murray-Darling Basin, one of the most important agricultural regions in the country. The economic impacts of the drought were severe, with agricultural output declining by approximately 50% in some years, leading to billions of dollars in economic losses (Smith, 2019). Many farmers were forced to shift to less water-intensive crops, abandon high-value crops like cotton and rice, or exit farming altogether, which profoundly affected local economies. Communities that had relied on agriculture as a primary source of income experienced decreased revenue, job losses, and shifts in local economic structures. In response, the Australian government invested in water infrastructure, improved water management practices, and introduced policies to increase resilience against future droughts.

In Africa, the Sahel region has long grappled with water scarcity, which has hindered agricultural development and exacerbated food insecurity. The Sahel, a semi-arid region stretching across multiple countries, experiences highly variable rainfall and frequent droughts that severely limit agricultural productivity (Anderson, 2018). The scarcity of water in this region has led to recurring food shortages, resulting in high food prices and increased reliance on imported food, which strains both government budgets and household finances. To address these challenges, countries in the Sahel have implemented sustainable water management practices, such as small-scale irrigation systems, rainwater harvesting, and agroforestry. These strategies aim to improve food security and reduce the economic impacts of water scarcity by increasing agricultural productivity and resilience to drought. Rainwater harvesting and agroforestry, for example, help conserve soil moisture, support crop growth, and provide additional sources of income, demonstrating how local adaptation measures can mitigate the adverse effects of water scarcity.

These case studies illustrate the profound economic impacts of water scarcity on agriculture and the broader economy. In regions prone to drought, the challenges of managing limited water resources are magnified, affecting not only agricultural productivity but also food security, employment, and economic stability. For areas that depend heavily on agriculture, prolonged droughts can disrupt local economies, inflate food prices, and increase poverty, creating a cycle of vulnerability that can be difficult to break. The economic toll of water scarcity often extends to national and global markets, as reduced agricultural output in major producing regions affects food availability and prices worldwide.

Addressing these challenges requires a combination of policy interventions, investments in water infrastructure, and the promotion of water-efficient practices. Governments and organizations play a crucial role in implementing policies that regulate water use, provide subsidies for water-saving technologies, and encourage sustainable water management practices. Investment in water infrastructure, such as dams, reservoirs, and water recycling facilities, can provide critical support during droughts, enabling agricultural systems to adapt to fluctuating water availability. Additionally, the adoption of water-efficient practices, including precision irrigation, crop rotation, and drought-resistant crop varieties, can help farmers reduce water usage and maintain productivity under water-limited conditions.

As climate change continues to exacerbate water shortages, the urgency for sustainable water management in agriculture is growing. Climate change is expected to increase the frequency and intensity of droughts, further straining water resources and agricultural systems worldwide. Developing adaptive strategies to cope with these changes is essential for sustaining agricultural production, protecting food security, and ensuring economic stability. By implementing water management practices that prioritize conservation, efficiency, and resilience, regions prone to drought can better prepare for the challenges posed by climate change, securing agricultural systems and livelihoods for the future.

Chapter 4

Chemical Costs and Emissions of Fertilizers and Pesticides

Fertilizers and pesticides have revolutionized modern agriculture, transforming food production and enabling unprecedented crop yields to meet the demands of a growing global population. These chemical inputs form the backbone of industrial farming, allowing for the intensive cultivation of staple crops and supporting large-scale monocultures. Synthetic fertilizers, rich in essential nutrients like nitrogen, phosphorus, and potassium, have dramatically increased soil productivity, while pesticides shield crops from the relentless threats posed by pests, diseases, and weeds. However, this reliance on chemical inputs comes at a steep cost. Beneath their undeniable contributions to food security lies a complex web of environmental, economic, and health consequences. From the greenhouse gases emitted during their production and use to the contamination of waterways, soil degradation, and biodiversity loss they leave in their wake, fertilizers and pesticides represent both a solution to food scarcity and a challenge to sustainability. This chapter delves into the dual role these chemicals play in agriculture, exploring their critical importance, hidden costs, and the urgent need for sustainable alternatives.

The Role of Synthetic Fertilizers and Pesticides in Food Production

Synthetic fertilizers and pesticides have become essential tools in modern agriculture, playing a critical role in increasing crop yields and enhancing food security. Synthetic fertilizers, primarily composed of nitrogen, phosphorus, and potassium (NPK), provide essential nutrients that support plant growth, increase agricultural productivity, and enable intensive farming practices that would otherwise be unsustainable. Nitrogen is particularly crucial, as it is a key component of chlorophyll, the molecule plants use in photosynthesis, and is involved in forming amino acids, the building blocks of proteins. Phosphorus aids in energy transfer within plants and root development, while potassium enhances drought tolerance and resistance to diseases (Brown & Lee, 2019). These fertilizers were instrumental during the Green Revolution, a period in the mid-20th century that saw the introduction of high-yield crop varieties, advanced farming techniques, and increased use of chemical inputs. This revolutionized food production in many parts of the world, significantly increasing yields and helping to prevent widespread food shortages, especially in densely populated regions like South Asia (Anderson, 2020). The application of synthetic fertilizers enabled countries to meet rising food demands, support growing populations, and improve nutritional outcomes.

However, the widespread use of synthetic fertilizers is also associated with serious environmental and health concerns. When applied to fields, these chemicals often leach into soils and waterways, leading to pollution and ecosystem imbalances. Nitrogen fertilizers, for example, can leach into groundwater as nitrate, contaminating drinking water sources. High nitrate levels in drinking water are linked to health risks, including methemoglobinemia, or "blue baby syndrome," which affects infants by reducing the blood's oxygen-carrying capacity (Johnson & Carter, 2017). In surface waters, nitrogen and phosphorus runoff contributes to nutrient pollution, which stimulates excessive algal growth. When these algae decompose, they deplete oxygen in the water, creating "dead zones" where aquatic life cannot survive. The Gulf of Mexico is one prominent example, where fertilizer runoff from the Mississippi River watershed has led to one of the largest dead zones in the world, with significant ecological and economic impacts on local fisheries (Smith, 2019).

Pesticides, which include insecticides, herbicides, and fungicides, are used to protect crops from pests, diseases, and weeds that can cause substantial yield losses. Herbicides like glyphosate help control weed growth, reducing competition for nutrients, light, and water and allowing crops to thrive with minimal interference. Glyphosate, one of the most widely used herbicides globally, has been credited with enabling no-till farming practices, which help reduce soil erosion and preserve soil health (Garcia, 2018). Insecticides target specific insect pests, reducing crop damage and loss, while fungicides protect plants from diseases caused by fungi and other pathogens. These chemicals have been highly effective in minimizing crop losses, contributing to steady food supplies, and supporting large-scale monoculture farming, where single crops are grown over extensive areas, maximizing efficiency and yield potential.

However, the intensive use of pesticides has raised significant concerns regarding their impact on the environment and human health. Many pesticides are non-selective, meaning they affect non-target species as well as the intended pests. For instance, insecticides can harm beneficial insects such as bees and butterflies, which are essential for pollination. The decline in pollinator populations, partly attributed to pesticide exposure, threatens the production of many crops that rely on these species, posing risks to both agricultural output and biodiversity (Johnson & Carter, 2017). Additionally, pesticides can persist in soils and water, accumulating over time and affecting ecosystems. Runoff from agricultural fields carries these chemicals into nearby rivers, lakes, and groundwater, potentially contaminating drinking water sources and harming aquatic life. Exposure to certain pesticides has been linked to adverse health effects in humans, including neurological disorders, endocrine disruption, and an increased risk of certain cancers, particularly for farmworkers and communities near agricultural areas (Smith, 2019).

Despite their benefits in boosting agricultural output, synthetic fertilizers and pesticides have also been linked to negative impacts on soil health, water quality, biodiversity, and human health. Overuse or improper application of these chemicals can disrupt soil ecosystems, harming beneficial organisms like earthworms, mycorrhizal fungi, and soil microbes that are essential for nutrient cycling, soil structure, and plant health. For example, earthworms, which aerate the soil and help decompose organic matter, can be killed by certain pesticides, leading to poorer soil health and reduced natural fertility (Garcia, 2018). Soil microbes, which play a vital role in breaking down organic material and releasing nutrients, can also be affected by chemical inputs, resulting in diminished soil biodiversity and nutrient imbalances. In the long term, this dependency on synthetic inputs can lead to a decline in soil fertility, creating a cycle where farmers must continually increase fertilizer applications to maintain yields.

The runoff of fertilizers into waterways is another serious environmental cost associated with synthetic inputs. Nitrogen and phosphorus runoff from agricultural fields contributes to harmful algal blooms in rivers, lakes, and coastal areas, resulting in hypoxic conditions where oxygen levels are too low to support fish and other marine life. These dead zones disrupt local fishing economies and damage biodiversity, affecting entire ecosystems. In addition to affecting marine life, nutrient pollution from agricultural runoff can compromise water quality, impacting recreational activities, tourism, and public health (Smith, 2019). Furthermore, nitrogen-based fertilizers are a source of nitrous oxide, a potent greenhouse gas that contributes to climate change. Agriculture is a major contributor to nitrous oxide emissions, and reducing fertilizer use is considered essential for lowering the sector's carbon footprint (Anderson, 2020).

These environmental and health impacts underscore the need for careful management and consideration of sustainable alternatives to reduce agriculture's reliance on synthetic inputs. Sustainable practices such as organic farming, crop rotation, and integrated pest management (IPM) offer viable solutions to minimize chemical use while maintaining productivity. Organic farming, which avoids synthetic chemicals in favor of natural fertilizers like compost and manure, promotes soil health and biodiversity by fostering a more balanced ecosystem. Crop rotation, which involves alternating crops in a field each season, can reduce pest and disease pressures naturally, decreasing the need for pesticides (Brown & Lee, 2019). IPM, which integrates multiple pest control methods, uses chemical pesticides only as a last resort, relying instead on biological controls, cultural practices, and mechanical methods to manage pest populations (Garcia, 2018).

In summary, while synthetic fertilizers and pesticides have been essential in supporting high-yield agriculture and meeting global food demands, their environmental and health costs highlight the challenges of chemical-intensive farming. By investing in sustainable practices, agriculture can reduce its reliance on synthetic inputs, protect ecosystems, and promote a healthier, more resilient food system for future generations.

Environmental and Economic Costs of Chemical Reliance

The environmental costs of relying on synthetic fertilizers and pesticides are substantial, impacting ecosystems, human health, and contributing significantly to climate change. The production of synthetic nitrogen fertilizers, for example, is an energy-intensive process dependent on the Haber-Bosch process, which combines nitrogen and hydrogen at high temperatures and pressures to create ammonia. This process relies heavily on fossil fuels, especially natural gas, resulting in considerable carbon dioxide emissions (Garcia, 2018). Globally, synthetic nitrogen production is responsible for a significant portion of agriculture's greenhouse gas emissions, contributing to the sector's overall environmental footprint. Additionally, when applied to soil, nitrogen fertilizers release nitrous oxide, a greenhouse gas with a global warming potential approximately 300 times that of carbon dioxide, making it a critical factor in agriculture's contribution to climate change (Anderson, 2020). These emissions have long-lasting effects on the global climate, highlighting the impact of synthetic fertilizers beyond the farm.

In addition to greenhouse gas emissions, the overuse of synthetic fertilizers has led to severe soil degradation. Excessive fertilizer application disrupts the natural balance of soil nutrients, often leading to nutrient imbalances that undermine soil productivity over time. For instance, repeated nitrogen application can lead to phosphorus and potassium deficiencies, requiring additional chemical inputs to maintain yields (Brown & Lee, 2019). Furthermore, soil organisms, including beneficial bacteria, fungi, and earthworms, which play essential roles in nutrient cycling and organic matter decomposition, are often negatively impacted by chemical residues. These organisms are crucial for maintaining healthy soil structure and fertility, and their decline weakens the soil's natural resilience, leading to a dependency on synthetic fertilizers to maintain productivity. This dependency not only increases costs for farmers but also reduces the soil's capacity to sequester carbon, further contributing to climate change.

The widespread use of pesticides has additional negative impacts on ecosystems and biodiversity. Pesticides, while effective at controlling specific pests, often harm non-target organisms, including pollinators like bees, butterflies, and beneficial insects that help maintain ecological balance. This loss of pollinators is particularly concerning, as many crops depend on these species for reproduction, and declines in pollinator populations have been linked to pesticide exposure. Pesticide runoff into nearby water bodies further exacerbates ecological damage, contaminating aquatic ecosystems and harming fish, amphibians, and other wildlife. Pesticide residues in soil also contribute to soil degradation, affecting soil microorganisms that support plant health and nutrient availability. Over time, this diminishes soil biodiversity, creating conditions that may require even greater pesticide use (Smith, 2019).

One of the unintended consequences of intensive pesticide use is the development of pesticide-resistant pest populations. When pests are exposed to pesticides repeatedly, they can develop genetic resistance, rendering the chemicals less effective over time. To combat this resistance, farmers are often forced to apply higher doses or switch to stronger, more toxic pesticides, which escalates both economic and environmental costs (Garcia, 2018). This cycle of resistance contributes to what is known as the "pesticide treadmill," where increasing reliance on chemical inputs leads to diminishing returns in pest control, higher costs, and escalating risks to both environmental and human health.

The economic costs associated with heavy reliance on synthetic chemicals are also substantial. Fertilizers and pesticides constitute a significant portion of production costs for farmers, and rising prices for these inputs can make it difficult for farmers to maintain profitability. This is particularly challenging for smallholder farmers in developing countries, where access to financial resources and subsidies is limited. For these farmers, the high costs of synthetic inputs can lead to financial strain, forcing them to take on debt or, in some cases, abandon farming altogether. This trend has serious implications for food security, as smallholder farmers are responsible for producing a significant share of the world's food supply, particularly in regions where subsistence farming is prevalent (Smith, 2019). The economic burden of synthetic inputs, combined with declining soil productivity, creates a challenging environment for sustainable farming, especially for small-scale operations.

In addition to direct costs for farmers, the environmental damage caused by chemical runoff into waterways necessitates costly clean-up and mitigation efforts. Fertilizer runoff into rivers, lakes, and coastal waters leads to nutrient pollution, causing harmful algal blooms and hypoxic zones where oxygen levels are too low to support marine life. These "dead zones" have severe economic impacts on fisheries and local economies that depend on healthy aquatic ecosystems. Addressing these issues often requires significant public investment in water treatment and restoration projects, costs that are typically borne by taxpayers rather than by the agricultural sector. These hidden environmental costs, coupled with the economic strain on farmers, highlight the need for a shift toward more sustainable practices that reduce reliance on synthetic inputs and promote long-term agricultural resilience.

In response to these environmental and economic challenges, sustainable agricultural practices are gaining attention as viable alternatives to chemical-intensive farming. Approaches such as organic farming, which relies on natural fertilizers like compost and manure, can help restore soil health and biodiversity. Crop rotation and cover cropping are additional strategies that enhance soil structure, prevent erosion, and reduce the need for chemical inputs by naturally managing soil fertility and pest populations (Brown & Lee, 2019). Integrated Pest Management (IPM) is another sustainable alternative, combining biological control, habitat manipulation, and targeted pesticide application to manage pests while minimizing environmental impacts (Garcia, 2018). By promoting practices that align with natural processes, sustainable agriculture aims to reduce the environmental footprint of food production, improve soil health, and lower the economic burden on farmers, fostering a more resilient agricultural system.

Sustainable Alternatives: Organic Farming, Integrated Pest Management

To address the environmental and economic costs of synthetic fertilizers and pesticides, sustainable alternatives such as organic farming and integrated pest management (IPM) are increasingly being adopted worldwide. Organic farming takes a holistic approach by eliminating synthetic inputs in favor of natural soil amendments, crop rotation, and biological pest control methods. Rather than focusing solely on crop yields, organic farming emphasizes soil health and biodiversity, which helps create resilient ecosystems and reduces the need for chemical fertilizers and pesticides (Johnson & Carter, 2017). Organic farmers use natural fertilizers, such as compost and manure, which enrich soil fertility without the pollution risks associated with synthetic fertilizers. By improving soil structure and organic matter content, these natural amendments help soils retain moisture and nutrients, making them more resilient to drought and extreme weather. Additionally, cover crops are used to prevent soil erosion, enhance nutrient cycling, and suppress weeds, all of which support soil health while minimizing the need for external inputs.

One of the significant benefits of organic farming is its role in mitigating climate change through carbon sequestration. Soils managed with organic practices tend to have higher levels of organic carbon, as organic matter from compost, manure, and crop residues is continuously added to the soil. This stored carbon not only enhances soil fertility but also reduces atmospheric carbon dioxide levels, contributing to global efforts to curb climate change (Johnson & Carter, 2017). Additionally, by avoiding synthetic fertilizers, organic farming reduces nitrous oxide emissions, a potent greenhouse gas released when nitrogen fertilizers break down in soil. This dual role of organic farming in both enhancing soil quality and reducing greenhouse gas emissions makes it an increasingly valuable strategy for sustainable agriculture.

Integrated Pest Management (IPM) is another sustainable alternative that combines biological, cultural, physical, and chemical control methods to manage pest populations effectively and reduce reliance on pesticides. Unlike conventional pest control, which typically relies on routine pesticide applications, IPM emphasizes preventive measures and regular monitoring of pest levels to determine the most appropriate and environmentally friendly control methods. Biological controls play a central role in IPM, with techniques such as the introduction of natural predators, parasites, or pathogens to manage pest populations without harming non-target species or beneficial insects. For example, ladybugs, which feed on aphids, can be introduced to protect crops from these pests naturally (Garcia, 2018). By supporting biodiversity, biological controls within IPM enhance the ecosystem's natural balance, minimizing pest outbreaks and reducing the need for chemical interventions.

Cultural practices are also essential components of IPM, helping to disrupt pest life cycles and reduce pest populations naturally. Crop rotation, for instance, prevents pests from establishing themselves in a particular field, as rotating crops changes the conditions and food sources available to them. Intercropping, or planting different crops together, can also deter pests, as diverse plantings make it harder for pests to locate and infest specific crops. Physical methods, such as traps and barriers, further contribute to IPM by physically preventing pests from reaching crops. When chemical controls are necessary, IPM emphasizes selective and targeted pesticide applications as a last resort. By carefully choosing the least harmful pesticides and applying them in minimal amounts, IPM reduces environmental impact and minimizes the risk of pests developing resistance to chemicals.

While organic farming and IPM offer promising alternatives to conventional farming methods, they also present challenges. Organic farming often requires higher labor inputs, particularly for activities like manual weeding, composting, and pest monitoring, which can increase production costs. Additionally, organic systems may experience lower yields compared to conventional systems, especially in the initial transition period when soil health is still recovering from prior chemical use. Similarly, IPM demands careful pest monitoring and planning, requiring farmers to understand and implement diverse pest control methods, which can be more time-consuming and require additional knowledge and training (Garcia, 2018). These challenges can make sustainable farming practices less accessible for small-scale farmers with limited resources or technical expertise.

Despite these challenges, the long-term benefits of sustainable practices, such as improved soil health, enhanced biodiversity, reduced greenhouse gas emissions, and lower environmental costs, make them essential components of a resilient agricultural system. Recognizing these benefits, governments and organizations around the world are increasingly supporting the transition to sustainable farming practices through subsidies, educational programs, and research into more efficient organic and IPM methods. Subsidies and financial incentives help offset the initial costs associated with transitioning to organic and IPM practices, making them more accessible to a broader range of farmers. Additionally, educational programs and extension services provide farmers with the knowledge and skills needed to implement these practices effectively, addressing the challenges of labor demands and technical complexity (Smith, 2019). Research and innovation in sustainable farming continue to improve these practices, exploring new organic fertilizers, natural pest control agents, and precision agriculture technologies that can enhance efficiency and reduce labor requirements.

By promoting organic farming and IPM, the agricultural sector can reduce its reliance on chemical inputs, protect ecosystems, and foster food security in an era of increasing environmental challenges. These sustainable practices offer pathways to mitigate the environmental and economic costs of conventional agriculture while building a foundation for resilient, productive, and ecologically sound food systems. As climate change, soil degradation, and biodiversity loss continue to threaten global food security, adopting sustainable practices becomes not only a choice but a necessity for the future of agriculture.

Chapter 5

Animal Agriculture and Emission Hotspots

Animal agriculture sits at the heart of the global food chain, fulfilling humanity's demand for protein while exacting an enormous environmental cost. Encompassing beef, poultry, pork, and seafood production, this sector is among the most resource-intensive and emission-heavy industries, contributing significantly to deforestation, water scarcity, and greenhouse gas emissions. Beef production, in particular, stands out as one of the largest environmental offenders, with its massive requirements for land, feed, and water, alongside the methane emissions generated by ruminant livestock. However, other forms of animal agriculture, including poultry, pork, and aquaculture, also come with considerable environmental trade-offs, from nutrient pollution in waterways to the overuse of antibiotics and growth hormones. As global demand for animal protein continues to rise, the environmental footprint of this sector expands, amplifying the urgency of rethinking agricultural practices. This chapter examines the emission hotspots within animal agriculture, explores the impacts of livestock farming on ecosystems and climate, and highlights the growing role of plant-based alternatives as a solution to the industry's pressing sustainability challenges.

Examining the Meat Industry: Beef, Poultry, Pork, and Seafood

Animal agriculture, encompassing the production of beef, poultry, pork, and seafood, is a primary driver of global greenhouse gas emissions and environmental degradation. These industries rely on intensive farming methods that demand vast resources, including land, water, and feed, leading to extensive environmental impacts. Beef production, particularly, stands out as one of the most resource-intensive agricultural practices. Cattle farming requires immense quantities of feed, water, and land for grazing, driving deforestation and biodiversity loss, especially in regions like the Amazon (Smith & Johnson, 2018). Furthermore, beef production generates significant greenhouse gases, particularly methane, due to enteric fermentation in ruminant animals, along with carbon dioxide emissions from the energy required for feed cultivation and transport. Research indicates that beef production emits approximately 27 kilograms of CO_2-equivalent per kilogram of meat, making it one of the most emission-intensive foods (Garcia, 2020).

While poultry and pork production have lower greenhouse gas emissions than beef, they still have considerable environmental consequences. Poultry farming uses less land and feed compared to cattle, but the industry requires substantial water, contributing to regional water scarcity. Additionally, poultry and pork farms produce large amounts of manure, leading to nitrogen pollution and nutrient overload in local ecosystems, which can trigger harmful algal blooms and create hypoxic (low-oxygen) zones detrimental to aquatic life (Lee et al., 2019). Managing waste from these farms is a major environmental concern, as nutrient runoff from manure can contaminate nearby waterways, compromising water quality. Furthermore, the widespread use of antibiotics and growth hormones in these sectors raises public health concerns, as it can accelerate antibiotic resistance and introduce hormones into the food supply, posing potential risks to human health (Thompson, 2018).

The seafood industry also presents unique environmental challenges. While wild-caught seafood tends to have a lower carbon footprint compared to terrestrial livestock, overfishing, bycatch (the capture of unintended species), and habitat destruction pose severe ecological threats (Brown & Harris, 2017). Overfishing can lead to the collapse of fish populations, diminishing biodiversity and disrupting ocean ecosystems that depend on healthy fish stocks. Although aquaculture, or fish farming, is an alternative to wild-capture fishing, it brings its own environmental issues, such as water pollution, disease spread, and the need for wild fish to produce feed, which can further strain marine resources. Aquaculture operations can cause nutrient pollution in local waters, threatening coastal ecosystems and contributing to the decline of wild fish populations.

The global demand for animal protein continues to rise, intensifying the environmental footprint of animal agriculture. As the industry expands, so do the pressures on ecosystems, from deforestation and water pollution to greenhouse gas emissions and habitat loss. Sustainable solutions and reforms in agricultural practices are increasingly essential to mitigate these environmental impacts and address the broader challenges posed by the growing demand for meat and seafood.

Methane Emissions and Environmental Impacts of Livestock

Methane emissions from livestock represent a critical environmental challenge in animal agriculture, largely due to methane's potency as a greenhouse gas. With a global warming potential approximately 28-36 times greater than carbon dioxide over a 100-year span, methane significantly accelerates climate change (Garcia, 2020). Ruminant animals like cattle and sheep are primary methane producers, as they release this gas through enteric fermentation—a digestive process unique to ruminants that generates methane as a byproduct. Research estimates that methane emissions from livestock comprise nearly 40% of all agricultural greenhouse gas emissions worldwide, with cattle alone responsible for the majority due to their sheer numbers and substantial methane output per animal (Smith & Johnson, 2018).

Beyond methane emissions, livestock farming drives other substantial environmental issues, including land degradation, deforestation, water pollution, and biodiversity loss. Large areas of land are often cleared to provide grazing space and cultivate feed crops, contributing to extensive deforestation in sensitive ecosystems, such as the Amazon rainforest. The clearance of these forests releases significant amounts of stored carbon dioxide, compounding greenhouse gas emissions while simultaneously diminishing biodiversity. Many plant and animal species lose their habitats, leading to disrupted ecosystems and increased risks of extinction (Brown & Harris, 2017).

Water use in livestock farming, especially in beef production, further intensifies environmental strain. Producing just one kilogram of beef can demand as much as 15,000 liters of water, factoring in the water needed for feed crops, drinking, and general upkeep of the animals (Lee et al., 2019). This high water demand places pressure on already scarce water resources, exacerbating water scarcity in regions heavily reliant on animal agriculture. Additionally, runoff from livestock operations can lead to severe water pollution. Manure from large-scale livestock farms is often produced in quantities far exceeding what surrounding land can absorb, leading to nutrient-rich runoff into nearby water bodies. This runoff introduces high levels of nitrogen and phosphorus into aquatic ecosystems, promoting algal blooms that deplete oxygen levels and create hypoxic zones, or 'dead zones,' where aquatic life struggles to survive (Thompson, 2018).

The environmental costs of livestock farming are broad and far-reaching. The sector impacts air quality through methane emissions, diminishes water quality due to nutrient runoff, leads to extensive land-use changes through deforestation, and contributes significantly to climate instability. Addressing these issues requires a shift toward more sustainable practices in animal agriculture, balancing the demand for animal products with the imperative to protect and preserve environmental health.

Plant-Based Alternatives: Costs, Carbon Savings, and Adoption Trends

In response to the significant environmental impacts of animal agriculture, plant-based alternatives are emerging as a sustainable and environmentally friendly solution. These alternatives, which include proteins derived from sources such as soy, peas, beans, and lentils, offer substantial reductions in carbon emissions compared to traditional animal-based proteins. Studies consistently show that plant-based foods produce considerably lower greenhouse gas emissions, require significantly less water, and occupy much less land than their animal-based counterparts (Garcia, 2020). For instance, producing one kilogram of plant-based protein generates only a fraction of the CO_2-equivalent emitted by beef, highlighting plant-based diets as a promising strategy for reducing the carbon footprint of agriculture.

However, while the environmental benefits of plant-based alternatives are clear, their production costs remain relatively high compared to conventional meat products. This is primarily due to the relatively young stage of the plant-based food industry and the lack of economies of scale. As the industry matures and consumer demand for plant-based products rises, production costs are expected to decrease, making these options more accessible. Additionally, plant-based foods offer significant environmental savings by reducing land and water use, minimizing pollution, and curbing greenhouse gas emissions (Smith & Johnson, 2018). Innovators like Beyond Meat and Impossible Foods are leading the way in creating plant-based products that closely mimic the taste, texture, and cooking experience of meat, helping to bridge the gap for consumers who seek sustainable yet satisfying dietary options.

Globally, the adoption of plant-based alternatives varies but is steadily increasing, particularly in high-income countries where environmental awareness, health consciousness, and ethical concerns over animal welfare drive dietary shifts. Millennials and Generation Z are at the forefront of this demand for plant-based options, motivated by concerns about health, animal welfare, and sustainability (Thompson, 2018). To support these shifts, governments and organizations are increasingly encouraging plant-based diets through updated dietary guidelines that emphasize reduced meat consumption. For example, the EAT-Lancet Commission's "planetary health diet" advocates for a diet rich in plant-based foods and limited in red meat, with the dual goal of promoting human health and environmental sustainability (Brown & Harris, 2017).

The rise of plant-based alternatives offers a promising path forward in addressing the environmental challenges posed by animal agriculture. By lowering methane emissions, reducing deforestation rates, and conserving water resources, plant-based diets could contribute substantially to environmental protection. The growing interest in these alternatives aligns with broader efforts to create a resilient and sustainable food system that supports both planetary and human health. As demand grows, continued research and innovation will be essential to enhance the affordability, variety, and accessibility of plant-based foods. Addressing challenges related to consumer acceptance and the need for widespread availability across different regions will be crucial in establishing plant-based diets as a mainstream solution for a more sustainable future.

Chapter 6

The Carbon Journey from Farm to Market

Transportation is the lifeline of the modern food system, connecting farms to consumers and enabling the global exchange of diverse food products. Yet, this essential process comes with a substantial environmental cost, as the movement of food across regions and continents contributes significantly to greenhouse gas (GHG) emissions. The concept of "food miles" highlights this impact, measuring the distance food travels and its corresponding carbon footprint. Whether it's exotic fruits flown thousands of miles or locally sourced vegetables transported a few towns away, every step in the journey from farm to plate plays a role in determining the sustainability of the food supply chain. As consumer preferences for variety, convenience, and year-round availability continue to grow, so too do the challenges of managing the carbon footprint of food transportation. Understanding the implications of food miles and exploring innovative solutions to reduce emissions are critical steps toward building a more sustainable and resilient food system.

Food Miles and the Impact of Transportation on Carbon Footprint

The concept of "food miles" describes the distance food travels from its source of production to the consumer's plate, a journey that plays a major role in determining the overall carbon footprint of food products. Transportation is a critical phase in the food supply chain, as it involves moving products across varying distances and often necessitates significant fuel and energy resources. This movement has a direct environmental impact, with studies indicating that the longer the transportation distance, the higher the greenhouse gas (GHG) emissions, particularly when fossil-fuel-based transportation methods such as trucking, shipping, or air transport are used (Weber & Matthews, 2008). Each method has a unique environmental footprint; for instance, air transport emits substantially more CO_2 per mile than sea or road transport, making it the most carbon-intensive mode of food transportation. Foods that are sensitive to freshness or are highly perishable—such as berries, seafood, or specialty herbs—are often flown across continents to reach markets quickly. This practice, while economically beneficial in meeting consumer demand, considerably increases the carbon emissions of these products relative to those produced and consumed locally (McKinnon, 2018).

Food miles are particularly relevant as consumer demand grows for out-of-season produce and exotic foods, which are not locally available and must be imported from regions with suitable growing conditions. Many out-of-season products originate from tropical or subtropical climates, which can be thousands of miles away from their eventual retail destinations. This trend has been intensified by the globalization of food supply chains, which enables the year-round availability of a diverse array of foods. While these supply chains satisfy consumer preferences for variety and convenience, they come with significant environmental costs, particularly due to the emissions generated during the lengthy transportation process. This has led to a broader recognition of the need to reduce food miles by prioritizing local food sources, seasonal availability, and regional farming to help mitigate these impacts. By reducing the distance food must travel, local sourcing minimizes transportation-related emissions, supports local economies, and reduces dependency on imported goods (Edwards-Jones et al., 2008).

Furthermore, reducing food miles aligns with broader sustainability goals. Sourcing food locally not only cuts down on emissions from transportation but also boosts local economies and promotes food security within communities. Small-scale farmers benefit directly from increased local demand, allowing for economic stability and encouraging sustainable agricultural practices. Regional food systems can also contribute to community resilience by reducing reliance on complex global supply chains, which are susceptible to disruptions due to factors like fuel price volatility, political instability, or climate change impacts. Additionally, seasonal consumption encourages sustainable food practices by reducing the need for energy-intensive storage or rapid transportation of products that would otherwise be unavailable due to regional climate limitations. Emphasizing local food consumption also helps address food waste, as shorter supply chains reduce the risk of spoilage that can occur during long-haul transportation and extensive handling (Pretty et al., 2005).

In response to these concerns, there is an increasing push for consumers to consider the carbon footprint of their food choices and prioritize foods with fewer food miles. Some cities and regions have begun developing infrastructure to support local food distribution, including farmers' markets, community-supported agriculture (CSA) programs, and food hubs. These efforts seek to make local foods more accessible and convenient, providing consumers with fresher products that have traveled shorter distances. In addition, educational campaigns and labeling initiatives have emerged to inform consumers about the origin and carbon footprint of the foods they purchase. Through such initiatives, consumers are better equipped to make environmentally responsible choices, further driving demand for low-emission food options and encouraging the growth of local food economies.

The carbon impact of food transportation is thus influenced by multiple factors, including transportation mode, distance traveled, and logistical efficiency. Reducing food miles by supporting local sourcing, adjusting dietary habits to align with seasonal availability, and choosing foods that do not require intensive transportation can help lower the overall carbon footprint of the food system. These actions reflect a growing commitment to sustainable food practices and acknowledge the essential role of food miles in achieving environmental and economic sustainability.

Differences Between Local and Global Food Distribution Chains

Local and global food distribution chains exhibit substantial differences in transportation requirements, infrastructure needs, and environmental impacts. Local food distribution generally involves moving products across shorter distances, with fewer intermediaries and a lower reliance on extensive transportation and storage. These shorter distances result in lower greenhouse gas (GHG) emissions due to reduced fuel consumption, making local food systems inherently more sustainable in terms of carbon footprint. Local food systems are often organized around regional networks that connect farmers, markets, and consumers within the same area, which minimizes the need for large-scale transportation infrastructure and reliance on fossil fuel-dependent logistics (King et al., 2010). The shorter supply chains in local food distribution allow for fresher products that require less refrigeration and preservation energy, which is particularly advantageous in regions with climates that support year-round local agriculture. Additionally, local products are often minimally packaged or sold directly to consumers, reducing the need for extended shelf-life packaging and conserving resources.

In contrast, global food distribution chains are built on complex networks that transport food across continents, involving multiple stages of handling, processing, and storage. Global distribution often necessitates refrigeration, significant packaging, and substantial fuel usage to maintain food quality over long distances. These requirements contribute to GHG emissions and increased energy consumption, particularly when perishable foods are transported by air or over long distances by road, both of which are more carbon-intensive methods of transportation (Garnett, 2011). While global food distribution allows consumers access to a wide variety of foods year-round, it typically leads to a higher carbon footprint within the food system due to the reliance on fossil fuels and the environmental costs associated with maintaining quality across long distances. Global distribution is essential for ensuring food security in regions where local agriculture cannot meet all food needs, but it comes with considerable environmental trade-offs.

The environmental drawbacks of global food chains have led to a growing interest in alternative distribution models that reduce environmental impact while supporting local economies. Community-supported agriculture (CSA) and farmers' markets have gained popularity as local distribution models that allow consumers access to fresh, locally produced foods while minimizing emissions and reducing the dependency on international supply chains. CSAs connect consumers directly with local farmers, who provide fresh, seasonal produce without the need for long-haul transportation or significant packaging. Farmers' markets similarly promote direct sales between producers and consumers, fostering relationships within the community and reducing the environmental costs associated with traditional retail models (Coleman-Jensen et al., 2019).

Another advantage of local food systems is their reduced susceptibility to disruptions in global supply chains. Climate change, economic instability, and geopolitical tensions can all impact global food supply chains, leading to shortages, price fluctuations, and compromised food security. Local food systems can offer more stability and resilience in the face of these challenges, as they rely less on extensive transportation and are supported by regional networks that are often more adaptable to local conditions. Additionally, by reducing the dependency on imported foods, local food distribution systems can help communities decrease their carbon footprint while promoting sustainable agricultural practices that benefit the local environment and economy.

In terms of consumer impact, the choice between local and global food sources often reflects a balance of environmental, economic, and cultural factors. While global food systems are essential for accessing foods not locally available, especially in regions with limited growing seasons or challenging climates, local food systems provide opportunities for consumers to lower their environmental impact, support local economies, and promote food sovereignty. For example, consumers in temperate climates who prioritize local food options may have fewer out-of-season foods available but will benefit from lower emissions and fresher products. Alternatively, those relying on global food systems can enjoy year-round variety but contribute to higher emissions associated with transport and storage. Consequently, the choice between local and global food distribution is influenced by multiple factors, including availability, sustainability goals, and consumer preferences for convenience or diversity in food options.

Overall, the structural and operational differences between local and global food distribution chains highlight the complex balance between environmental sustainability and food accessibility. As awareness of the environmental impacts of food transportation grows, more consumers and businesses are exploring ways to reduce the carbon footprint of food supply chains by choosing local products, supporting sustainable agricultural practices, and investing in low-emission distribution models.

Innovations in Reducing Transportation Costs and Emissions

As the food industry intensifies its efforts to combat climate change, a variety of innovations are emerging to help reduce the carbon footprint associated with food transportation. Advances in technology, such as the development of fuel-efficient vehicles, alternative energy sources, and enhanced logistical systems, play a critical role in these efforts. For example, the adoption of electric and hybrid trucks in urban areas is helping reduce diesel emissions, particularly for short-haul transport that connects farms to nearby markets. These vehicles produce fewer emissions and are well-suited to frequent, short-distance trips, which are common in urban food distribution networks (Smith et al., 2020). Additionally, some companies are exploring biofuels and renewable energy options to power their transportation fleets, reducing reliance on fossil fuels and contributing to long-term sustainability goals.

Beyond improvements in vehicle technology, operational innovations in logistics and supply chain management are also proving effective in lowering emissions. Optimized routing, shipment consolidation, and collaborative distribution systems allow food distributors to minimize the number of trips and streamline their operations. For instance, combining multiple shipments into a single journey reduces the need for separate trips and decreases fuel consumption. Data analytics plays a significant role in optimizing routing by analyzing traffic patterns, delivery times, and vehicle capacity to select the most efficient routes (McKinnon, 2018). In urban areas, some companies are implementing "last-mile" solutions—using electric bicycles and drones to complete deliveries—thereby cutting down on emissions for short-distance, local transportation (Edwards-Jones et al., 2008). These last-mile innovations are particularly beneficial for densely populated areas where conventional vehicles face traffic congestion, resulting in higher emissions and delivery inefficiencies.

Packaging innovations further support emission reductions within food supply chains. Lightweight materials and compostable packaging not only reduce the energy required for transport but also lessen the environmental impact associated with packaging disposal. Lighter packaging materials enable vehicles to carry more products with less fuel, which is especially valuable for long-haul transportation. Compostable packaging aligns with sustainability goals by minimizing waste, thereby reducing landfill contributions and pollution from plastic waste. In combination, these advancements in packaging reduce both the carbon footprint of transporting food and the overall environmental impact of the supply chain.

Looking toward the future, sustainable food transportation is likely to rely on a blend of local food systems and technological advancements that promote energy efficiency and reduce emissions. The growing awareness of "food miles" and the impact of food transportation on climate change is motivating consumers, producers, and retailers alike to seek more sustainable food distribution practices. This shift includes an emphasis on locally sourced foods that can reduce the need for extensive transportation and minimize the associated emissions. In cases where local food is not available or feasible, advances in transportation technology and logistics can help mitigate the environmental costs of longer-distance food transport.

This transition toward lower-carbon food systems will require collaboration across multiple sectors, including agriculture, transportation, technology, and policy. Government incentives and regulations can encourage the adoption of eco-friendly technologies, such as electric vehicles and biofuels, while private companies can invest in research and development to continuously improve the sustainability of food transportation. By combining technological and operational innovations with support for local food systems, the food industry can make substantial progress in reducing its transportation-related emissions and building a more sustainable, resilient supply chain. This approach represents a comprehensive strategy that addresses the climate impacts of food transportation while meeting consumer demand for fresh, diverse, and accessible food options.

Chapter 7

Processing and Packaging
Hidden Carbon Costs

The journey food takes from farms to consumers' plates—whether across town or across the globe—is a critical yet often overlooked contributor to the environmental footprint of the food system. Transportation is an essential link in the food supply chain, enabling the delivery of fresh produce, meats, and packaged goods to global markets. However, the carbon costs of this journey are significant, with transportation emissions adding substantially to the food system's overall greenhouse gas (GHG) output. The concept of "food miles" has become a vital measure of these emissions, highlighting the distance food travels and the environmental costs of long-haul transportation methods like air freight and refrigerated trucking. As global demand grows for year-round availability of out-of-season produce and exotic foods, the reliance on energy-intensive transportation systems has surged, amplifying the food system's carbon footprint. This chapter delves into the environmental costs of food transportation, the contrasts between local and global food chains, and the innovations aimed at reducing emissions and building more sustainable distribution networks.

Energy Use in Food Processing and Packaging

The processing and packaging stages of the food supply chain are highly energy-intensive, often requiring large amounts of electricity, fuel, and water to transform raw ingredients into finished products, maintain their freshness, and prepare them for distribution. Each stage, from initial processing to final packaging, contributes to the overall carbon footprint of food products. Processing methods—such as heating, drying, freezing, milling, pasteurization, and sterilization—consume substantial amounts of energy, contributing to the environmental impact of food production. These methods are necessary to ensure food safety, extend shelf life, and create appealing textures and flavors, yet they account for a significant portion of the emissions produced throughout the food supply chain. Research estimates that food processing and packaging can account for up to 30% of the total energy consumption within the food supply chain, with much of this energy derived from fossil fuel sources that release greenhouse gases (GHGs) into the atmosphere (Garnett, 2011).

Industrial-scale food processing facilities, designed to operate continuously to meet demand, further amplify energy requirements. These facilities typically run 24/7 and use extensive systems for heating, cooling, drying, and mechanical processing, which ensures production efficiency but increases energy use and emissions. For instance, drying and freezing require high energy input, with drying alone accounting for up to 15% of industrial energy use in the food sector (Mujumdar & Law, 2010). Additionally, mechanical processing steps such as grinding, cutting, and emulsifying, though often less energy-intensive than thermal methods, still contribute to the overall energy footprint due to the scale at which they operate in industrial settings. To maintain efficient operations, these facilities rely on continuous fuel and electricity consumption, underlining the need for more energy-efficient technologies and practices to reduce the associated environmental impacts.

The packaging process further contributes to the energy demands and carbon footprint of the food industry. Packaging materials like plastic, glass, aluminum, and cardboard require substantial energy to produce, particularly through processes such as plastic extrusion, metal smelting, and glass molding. Each material has a distinct environmental profile; for example, aluminum is highly durable and recyclable but has one of the most energy-intensive production processes, involving significant electricity for smelting and emitting greenhouse gases such as carbon dioxide and perfluorocarbons (PFCs) during its production (Van der Harst & Potting, 2013). Plastic, while lightweight and versatile, relies on petroleum, and its production emits a considerable amount of CO_2, contributing to the overall carbon footprint of packaged goods (Barlow & Morgan, 2013). Glass, although fully recyclable, is heavy and requires substantial energy for production and transportation, adding to its carbon footprint.

The energy impact of packaging extends beyond material production, as many packaged food items require refrigeration or freezing to maintain quality from production to consumption. Cold storage, whether in industrial freezers or during transportation, consumes large amounts of energy, adding to the carbon cost of processed foods, especially for perishable items like meat, dairy, and frozen goods (Heller & Keoleian, 2015). The need for refrigeration not only increases emissions but also poses logistical challenges, as products must be kept within specific temperature ranges throughout the supply chain to avoid spoilage. This dependence on cold storage highlights the environmental burden of maintaining food freshness and underscores the importance of improving energy efficiency within refrigeration systems.

The total energy footprint of processing and packaging reveals the hidden environmental costs embedded in food production, transportation, and storage. These stages, while essential to meet consumer demand for safe and convenient foods, contribute heavily to GHG emissions, emphasizing the need for innovative solutions to reduce carbon emissions across the food industry. Increased investment in energy-efficient processing equipment, alternative packaging materials, and renewable energy sources can help mitigate these impacts. Strategies like optimizing heat recovery, improving insulation in refrigeration units, and developing less energy-intensive processing methods hold promise for reducing the carbon footprint of food processing and packaging.

Addressing the energy demands associated with these stages is essential for creating a more sustainable food system. Implementing energy-efficient technologies, reducing the use of single-use packaging, and embracing recyclable or biodegradable materials can collectively lower the environmental burden of food production. Such improvements not only align with sustainability goals but also respond to increasing consumer awareness and demand for eco-friendly food choices.

The Environmental Cost of Packaging Materials

Packaging materials play a significant role in the environmental footprint of food products due to their resource-intensive production, extended use, and challenging disposal requirements. Traditional packaging materials, including plastics, metals, glass, and cardboard, each contribute to environmental degradation in unique ways, primarily due to their dependence on non-renewable resources, energy-demanding manufacturing processes, and persistent waste management issues. Plastic, for instance, is one of the most prevalent materials in food packaging, valued for its versatility, durability, and low production costs. However, plastics are derived from petroleum, and the extraction and production of plastic emits significant amounts of CO_2, directly contributing to global warming (Heller & Keoleian, 2015). The environmental risks associated with plastic packaging do not end at production; plastic waste is a severe ecological hazard. It accumulates in landfills, oceans, and other ecosystems, where it can persist for centuries, releasing toxins into soil and water and posing a severe threat to marine and terrestrial wildlife. As plastic waste breaks down into microplastics, it enters the food chain, leading to potential health impacts for both animals and humans.

Metal packaging, commonly made from aluminum and steel, is also widely used in food preservation and distribution but has its own set of environmental consequences. The extraction and processing of metals are highly energy-intensive and result in substantial carbon emissions. Aluminum, in particular, requires significant electricity for production, especially during the smelting phase, which emits perfluorocarbon (PFC)—a potent greenhouse gas with a global warming potential much higher than CO_2 (Van der Harst & Potting, 2013). While metals like aluminum and steel are recyclable, the recycling rate for these materials in food packaging remains lower than desired. Reprocessing metals also requires a considerable amount of energy, although recycling aluminum, for instance, uses only 5% of the energy needed for primary aluminum production. Nonetheless, the total energy demand associated with metal packaging, coupled with the challenges in recycling infrastructure, contributes to the carbon footprint of packaged goods.

Glass packaging, often seen as a more sustainable alternative due to its inert and fully recyclable nature, also has notable environmental drawbacks. Glass is a heavy material, and the high temperatures required for glass production and molding involve intensive heat sources, usually derived from fossil fuels. These energy demands contribute significantly to GHG emissions. Moreover, the weight of glass adds to transportation emissions, as heavier loads require more fuel during transit. Although glass is theoretically 100% recyclable without loss of quality, the energy required to collect, transport, and recycle glass further contributes to its environmental impact. The carbon footprint of glass packaging, therefore, is not insignificant and is compounded by the emissions generated during both its production and transportation.

The waste management and disposal of packaging materials introduce further environmental challenges. Despite improvements in recycling infrastructure, a significant proportion of food packaging still ends up in landfills, where it can take years or even centuries to decompose, depending on the material. Plastic packaging is particularly problematic, as it does not biodegrade but instead photodegrades, breaking down into smaller fragments over time. These microplastics pose ecological hazards, as they are ingested by marine organisms, ultimately entering the food chain. Non-recyclable materials, or those contaminated with food residues, exacerbate the waste problem by clogging recycling streams and increasing landfill volume. The disposal process for packaging also releases greenhouse gases, including methane from decomposing organic waste and CO_2 from incineration. For instance, incinerating plastic waste generates CO_2 and other toxic emissions, such as dioxins, which pose health risks and contribute to air pollution. Methane emissions from landfills, in particular, have a global warming potential 25 times that of CO_2, emphasizing the climate impact of improper packaging disposal.

These environmental costs highlight the pressing need for the food industry to adopt sustainable packaging solutions that not only reduce waste but also lower the overall carbon footprint of food products. Sustainable alternatives, including biodegradable and compostable materials, as well as innovations like reusable packaging systems, offer potential solutions to mitigate the impact of traditional packaging. By reducing reliance on non-renewable materials, investing in eco-friendly production methods, and supporting waste management innovations, the food industry can address the hidden carbon costs of packaging.

Emerging Solutions: Minimalist Packaging, Edible Films, Biodegradable Options

In response to the significant environmental impact of traditional packaging, the food industry is actively exploring innovative solutions that reduce waste, decrease carbon emissions, and support a sustainable food system. One prominent approach is minimalist packaging, which focuses on minimizing the amount of material used by optimizing packaging shapes, sizes, and weights. This strategy not only conserves resources but also lowers transportation emissions, as more compact and lighter packaging allows for efficient storage, packing, and shipment (Robertson, 2012). Minimalist packaging is particularly effective for items with longer shelf lives, as it eliminates unnecessary materials without compromising food safety or product quality. By reducing the overall weight of packaged goods, minimalist packaging contributes to a decrease in fuel consumption during transportation, ultimately helping to cut down on greenhouse gas emissions.

Edible films are another promising development in sustainable packaging. Made from natural substances like proteins, polysaccharides, or lipids, edible films can be consumed along with the food product, thereby eliminating waste entirely. These films are especially beneficial for fresh produce and perishable items, as they provide a thin, protective barrier that helps to extend shelf life without generating plastic waste. Research indicates that edible films made from biodegradable materials, such as chitosan—a polysaccharide derived from crustacean shells—are effective at slowing moisture loss and microbial growth in fruits and vegetables, reducing the need for plastic wraps and synthetic preservatives (Krochta & Mulder-Johnston, 1997). While still under development, edible films have the potential to replace single-use plastics in specific food categories, offering a waste-free and environmentally friendly alternative that aligns with sustainability goals. By preventing food spoilage and offering biodegradable properties, these films support a closed-loop system, where packaging materials are either safely consumed or naturally decompose.

Biodegradable and compostable packaging solutions are also gaining traction as the demand for environmentally conscious alternatives increases. Unlike conventional plastics, which can persist in the environment for centuries, biodegradable materials are designed to break down naturally under specific composting conditions, resulting in minimal environmental impact. Compostable packaging, typically derived from plant-based materials like cornstarch or cellulose, decomposes into organic matter that can be used to enrich soil, aligning with the principles of a circular economy (Song et al., 2009). As the composting infrastructure continues to improve, compostable packaging is becoming a practical alternative for various food products, especially for single-use items. These materials are increasingly utilized in the packaging of fresh produce, takeout containers, and even cutlery, helping to reduce the amount of waste sent to landfills. However, successful adoption relies on proper disposal practices, as compostable items must be placed in appropriate composting facilities to decompose fully. Mismanagement or disposal in non-compostable waste streams can result in these materials being treated as conventional waste, which diminishes their environmental benefits.

The adoption of sustainable packaging solutions reflects a larger shift within the food industry to address the hidden carbon costs associated with food processing and packaging. As consumer awareness of environmental issues grows, food companies are responding by investing in research and development aimed at eco-friendly packaging options that align with global climate objectives. Many companies are experimenting with innovative materials, including mushroom-based packaging, which decomposes within weeks, and seaweed-derived materials that are entirely edible and dissolve in water. These efforts are not only reducing waste but also decreasing the industry's reliance on fossil fuel-based plastics and helping to establish more sustainable production and disposal methods.

In addition, companies are increasingly exploring partnerships with recycling and composting facilities to ensure that their biodegradable and compostable packaging solutions are properly processed. This industry-wide commitment to sustainability has also spurred the development of labeling standards that educate consumers on proper disposal practices, making it easier to distinguish between compostable, recyclable, and non-recyclable materials. For instance, the How2Recycle label provides consumers with clear information on whether a package is compostable, recyclable, or requires special disposal methods, supporting informed consumer choices and improving overall waste management.

The transition to sustainable packaging solutions is part of a larger, systemic change within the food industry as it adapts to meet both environmental and consumer demands. By embracing minimalist designs, investing in edible and compostable materials, and partnering with waste management facilities, the food industry is working to reduce its environmental footprint while supporting a more sustainable future. As these solutions become more accessible, they offer an opportunity to reshape the way food products are packaged and consumed, contributing to a cleaner, healthier environment for future generations.

Chapter 8

Economic Costs of Homegrown Foods

The economic costs associated with growing food locally, either through home gardening or community-supported agriculture (CSA), generally begin with an initial investment in essential resources. These initial costs can include purchasing seeds, soil amendments, gardening tools, and potentially more substantial expenses like raised garden beds or irrigation systems for larger plots. For home gardeners, soil quality is often a top priority, requiring amendments such as compost, organic fertilizers, or pH balancers to ensure that plants have the nutrients they need to thrive. Likewise, depending on local weather conditions and crop choices, some gardeners may need to install drip irrigation or other watering systems, adding to the setup costs. While these start-up investments can be significant, studies indicate that they are typically recouped over time as households save money by growing their own produce and reducing grocery bills (Johnson et al., 2015).

Beyond the initial setup, maintenance costs contribute to the ongoing investment required for successful home gardening. Regular costs include water usage, organic fertilizers, mulch, and pest control, which are necessary to maintain plant health and yield. Water, in particular, can be a notable expense, especially in arid regions or during dry seasons.

Many home gardeners incorporate sustainable practices, such as rainwater harvesting or mulching, to conserve water and reduce costs. Organic pest control methods, like neem oil, insecticidal soaps, or companion planting, can also add to the budget, but they are preferred by gardeners aiming to avoid synthetic chemicals. Despite these ongoing expenses, homegrown food can offer considerable long-term savings, particularly for high-yield crops such as tomatoes, cucumbers, herbs, and peppers, which can be relatively costly when purchased frequently from grocery stores. Studies show that well-maintained home gardens can save households several hundred dollars annually, making them an economically viable and sustainable source of fresh produce (Davis, 2016). Additionally, many gardeners find that growing their own vegetables improves the quality and taste of their food, providing added value beyond mere cost savings.

Community-supported agriculture (CSA) programs offer a different model for sourcing fresh produce locally and can be a cost-effective alternative to store-bought produce, particularly for households interested in supporting local farms. By purchasing a seasonal share from a nearby farm, consumers make an upfront investment that helps cover the farm's operational costs and, in return, receive weekly or biweekly boxes of seasonal fruits and vegetables. Although the upfront cost of a CSA share may initially seem high, it typically yields a large quantity of produce over the growing season, often surpassing the amount consumers could buy individually at retail prices. Studies indicate that CSA shares can offer substantial savings for consumers, as they receive diverse, farm-fresh produce at a lower overall cost than similar items from grocery stores, where prices are typically marked up to cover packaging, transportation, and retail expenses (Galt et al., 2019).

Both home gardening and CSA models provide consumers with the opportunity to invest in fresher, more affordable produce while reducing reliance on store-bought items. These models also present indirect economic benefits. Home gardening and CSA participation help localize food production, which supports regional economies by reducing dependence on imported foods and retaining more money within the community. Furthermore, locally grown produce often reaches consumers more quickly, preserving its nutritional value and reducing spoilage, which can lead to additional savings by minimizing food waste.

For households that grow their own food, the ability to preserve excess harvests—by canning, freezing, or drying—extends the economic and nutritional benefits of homegrown food year-round, providing another layer of long-term savings.

Ultimately, both home gardening and CSAs represent valuable pathways for households seeking more control over their food sources, better access to fresh produce, and meaningful economic savings. While initial investments may be a barrier for some, the long-term benefits often outweigh the costs, making these options attractive for consumers interested in sustainable and affordable food alternatives. Additionally, the social and environmental benefits of supporting local agriculture, reducing food miles, and promoting food self-sufficiency contribute to the growing popularity of these models as consumers become increasingly aware of the broader impacts of their food choices.

Environmental Impact of Local Growing: Reduced Food Miles and Carbon Savings

The environmental impact of locally grown food is evident in the significant reduction of food miles, which refers to the distance food travels from its source to the consumer. Locally sourced produce minimizes the need for extensive transportation, which in turn reduces emissions and decreases the overall carbon footprint of food products. Research indicates that food transportation—particularly by energy-intensive methods like air freight and long-haul trucking—contributes substantially to greenhouse gas (GHG) emissions. For example, fruits and vegetables imported from distant regions often require considerable fossil fuel usage, resulting in higher carbon emissions. Locally grown produce, such as vegetables cultivated in a community garden or backyard, avoids these fuel-intensive journeys, thus contributing to lower carbon emissions and energy savings (Weber & Matthews, 2008). This benefit is amplified when locally grown produce is also consumed seasonally, as seasonal foods are typically grown under natural conditions without the need for artificial climate control or preservation, both of which add to a food product's carbon footprint.

Seasonal, local produce aligns closely with environmental sustainability by reducing the reliance on energy-intensive preservation and storage methods, such as refrigeration, which is often essential for imported foods to maintain freshness during transport. Refrigeration alone accounts for a significant portion of the energy used in food distribution, particularly for perishable items. Locally grown, seasonal foods require less refrigeration, as they can often be consumed soon after harvest, eliminating the need for extended cold storage. This approach minimizes energy usage and further reduces GHG emissions associated with food preservation (Edwards-Jones et al., 2008).

In addition to reduced transportation and refrigeration demands, local and seasonal produce often requires fewer chemical inputs, such as pesticides and synthetic fertilizers, which are commonly used in large-scale agriculture to optimize yields and maintain quality during lengthy distribution. When crops are grown in their natural season and within their native or compatible climates, they generally thrive with less need for chemical intervention. This not only enhances soil and water health but also reduces the carbon footprint associated with the production, transportation, and application of these chemicals. Additionally, fewer chemical inputs mean reduced risks of pesticide and fertilizer runoff, which can contaminate local water sources and disrupt surrounding ecosystems. In contrast, large-scale farms often rely heavily on these inputs to maintain productivity, which indirectly contributes to higher GHG emissions due to the energy required in chemical production and distribution (Garnett, 2011).

The benefits of local and seasonal food systems extend beyond immediate environmental advantages, as they also support sustainable agricultural practices that contribute to long-term ecological health. Smaller-scale local farms tend to practice more sustainable farming techniques, such as crop rotation, organic farming, and conservation tillage, which promote soil fertility, reduce erosion, and enhance biodiversity. By fostering these practices, local food systems create a cycle that not only lowers emissions but also builds resilience against the environmental stresses associated with industrial agriculture. These systems help maintain healthy ecosystems and create a more sustainable model for food production, which is increasingly important as global food demands rise.

Overall, the environmental benefits of locally grown and seasonally aligned food are multi-faceted, providing a valuable alternative to the high-emission, energy-intensive processes associated with global food supply chains. By choosing local and seasonal produce, consumers can support reduced GHG emissions, lower energy consumption, and sustainable agricultural practices, ultimately contributing to a smaller carbon footprint and a more resilient food system.

Imported Foods: High Transportation Costs and Trade's Impact on Prices

Imported foods offer consumers a diverse range of produce and year-round availability, but they generally come with higher transportation and refrigeration costs due to the distances involved in international trade. The carbon footprint of imported foods is often substantially larger than that of locally sourced produce, as these items frequently require energy-intensive transportation methods like air freight or long-haul trucking to ensure freshness over extended distances. Foods such as avocados from Mexico, bananas from Central America, and tomatoes from Spain, for example, must often be refrigerated or shipped in climate-controlled containers to prevent spoilage during transit. This need for refrigeration, especially over extended periods, significantly raises the carbon footprint of imported foods due to the additional energy required, which directly contributes to greenhouse gas (GHG) emissions (McKinnon, 2018). The reliance on long-distance transport and cold storage solutions underscores the environmental costs associated with maintaining the global food supply chain, particularly for perishable goods that require careful handling to meet quality standards.

The refrigeration required for these foods is a major contributor to GHG emissions, as it consumes substantial amounts of energy, often from non-renewable sources. Cold storage not only keeps produce fresh but also extends shelf life, allowing foods to travel further distances; however, this energy-intensive process adds significantly to the overall environmental impact. Refrigeration systems are typically powered by electricity generated from fossil fuels, and they often rely on hydrofluorocarbon (HFC) refrigerants, which are highly potent greenhouse gases. According to research, emissions associated with refrigeration can constitute up to 15% of the total carbon footprint of certain imported foods, illustrating the hidden environmental costs associated with global food distribution (Heller & Keoleian, 2015). This dependency on refrigeration and energy-intensive transport highlights the carbon trade-offs inherent in the availability of imported foods, especially when compared to local alternatives that do not require such extensive storage and handling.

In addition to environmental costs, international trade in food introduces economic volatility that can influence the prices of imported items. Factors such as fluctuating fuel prices, tariffs, and trade policies play a significant role in shaping the price of imported foods, creating an unpredictable pricing structure that is influenced by global market forces. For instance, rising oil prices can directly increase the cost of transporting foods across long distances, which may then be passed on to consumers. Similarly, tariffs imposed on imported goods can raise prices, affecting consumer access to certain products. These costs fluctuate based on political and economic conditions, and they reflect the vulnerabilities of a global food system that relies heavily on complex logistics and fuel-dependent transport (Coleman-Jensen et al., 2019). As a result, the costs associated with importing foods can often be higher than those of locally sourced items, particularly in times of economic or political instability, further highlighting the economic trade-offs of choosing imported foods.

While international trade provides access to foods that cannot be locally grown due to climate or soil limitations, it also exposes consumers to an unpredictable price structure. Imported produce offers year-round access to popular fruits and vegetables, enabling consumers in colder climates to enjoy tropical produce, but these benefits come at a higher cost, both environmentally and economically. Market fluctuations and the complex web of international logistics mean that the price of imported foods can vary widely, depending on fuel costs, tariffs, and even seasonal demand. In comparison, locally sourced foods often have more stable pricing and require fewer resources to reach consumers, aligning with sustainable food practices and potentially offering a more cost-effective option.

The higher transportation and refrigeration costs of imported foods, combined with market volatility, underscore the trade-offs between convenience and sustainability. While consumers benefit from year-round variety and access to out-of-season produce, these choices come with increased environmental and economic costs. The reliance on fossil fuels for transportation and the energy demands of refrigeration contribute to the global carbon footprint, prompting a need for greater awareness and consideration of the impacts of food choices. As the climate impact of imported foods becomes more apparent, consumers may increasingly weigh these trade-offs when deciding between imported and locally grown options, balancing variety with sustainability in their food choices.

Case Studies: Carbon and Cost Comparisons of Imported vs. Local Foods

The carbon footprint of imported avocados is considerably higher than that of homegrown alternatives due to the extensive transportation, refrigeration, and agricultural practices required to meet global demand. Imported avocados, largely sourced from Mexico and Central America, often travel thousands of miles by truck or air freight before reaching consumers in other parts of the world, particularly in North America and Europe. Transporting avocados over such distances involves significant fuel consumption, and each stage of the journey—whether by air or road—contributes to greenhouse gas (GHG) emissions. Air freight, in particular, produces more CO_2 per mile than other modes of transport, and avocados frequently require air shipment to maintain their freshness due to their high perishability (Heller & Keoleian, 2015). Consequently, the carbon footprint of imported avocados is markedly high compared to foods that can be grown and consumed locally.

In addition to transportation, the refrigeration required to preserve avocados during transit adds to their environmental impact. Avocados are typically harvested before they fully ripen to extend their shelf life during transportation, which necessitates the use of climate-controlled containers and cold storage facilities. Refrigeration is an energy-intensive process that contributes significantly to the overall carbon emissions associated with avocados. A study by the Carbon Trust estimated that a single avocado imported to the United Kingdom could have a carbon footprint of up to 0.5 kg of CO_2-equivalent, with transportation and refrigeration accounting for a large portion of this impact. This figure varies depending on the distance traveled and the mode of transport, with air-freighted avocados producing the highest emissions per kilogram (Carbon Trust, 2019).

By contrast, homegrown avocados have a much smaller carbon footprint, as they do not require long-distance transportation or cold storage. When grown in suitable climates—such as in California or parts of Florida in the United States—avocados can be cultivated with minimal external inputs and are generally harvested closer to their optimal ripeness, reducing the need for extended refrigeration. The carbon emissions for homegrown avocados are largely limited to local transport (if necessary) and routine agricultural practices, which are typically far less energy-intensive than the logistics involved in global distribution. Homegrown avocados also allow for more sustainable agricultural practices, as small-scale growers often use fewer chemical inputs and more sustainable soil management techniques, further reducing the environmental impact.

This stark contrast between imported and locally grown avocados underscores the impact of food miles and transportation on a product's carbon footprint. Choosing homegrown avocados, when available, can significantly reduce one's carbon footprint by cutting out the energy-intensive transportation and refrigeration processes associated with imported produce. Moreover, supporting local avocado production contributes to regional agricultural resilience, reduces dependency on fossil fuel-based logistics, and promotes environmentally friendly food choices.

Challenges of Homegrown Foods: Navigating Availability, Seasonality, and Resource Demands

Despite the benefits of homegrown foods, several challenges limit their widespread adoption, particularly for certain crops and in specific regions. One of the primary challenges is the limited availability and seasonality of local foods. In many areas, especially those with colder climates, short growing seasons restrict the types of produce that can be cultivated year-round. Regions with harsh winters, for instance, are unable to grow fresh produce during colder months without advanced infrastructure like greenhouses, which can be costly to build and maintain. Additionally, certain fruits, such as bananas and avocados, thrive only in specific tropical or subtropical climates and are not viable for growth in many local environments, creating a reliance on imports to satisfy consumer demand for these foods. As a result, local food production is often limited to crops that are well-suited to the region's climate and seasonal conditions, which can restrict variety and year-round availability (Galt et al., 2019).

Another challenge lies in the time, effort, and knowledge required for effective home gardening and community-supported agriculture (CSA) participation. Home gardening demands a commitment to regular planting, watering, pruning, and harvesting, which can be overwhelming for individuals who have limited time or experience in horticulture. CSA programs, while providing access to local produce, require consumers to pick up weekly or biweekly shares and may involve volunteering or working on the farm, which can be challenging for those with busy schedules. A lack of gardening knowledge—such as how to select appropriate crops, manage soil health, or address pests without chemicals—can also be a barrier for new gardeners. While resources like gardening workshops or online tutorials are available, the learning curve for successful gardening is often steep, making it challenging for beginners to sustain the practice over time (Galt et al., 2019).

Resource requirements for successful gardening present further obstacles, particularly in regions facing water scarcity or poor soil quality. Water, for instance, is essential for crop growth, but many areas experience limited rainfall or face restrictions on water usage during droughts. In arid regions, gardening can be water-intensive, requiring efficient irrigation systems like drip irrigation to minimize waste, which adds to initial setup costs. Soil quality is another critical factor, as nutrient-rich, well-draining soil is essential for healthy plant growth. In areas with degraded or compacted soils, gardeners may need to invest in soil amendments, such as compost, organic fertilizers, or raised beds filled with high-quality soil. Additionally, pest control can be resource-intensive, as some plants are vulnerable to insects, fungi, and other threats that can damage crops if not managed properly. Organic or eco-friendly pest control options, while sustainable, can be more labor-intensive and costly than conventional pesticides, adding another layer of challenge (Davis, 2016).

Community-supported agriculture programs also face limitations that can impact their effectiveness and accessibility. CSA shares typically consist of seasonal produce, which means consumers may not receive the same variety they might expect at a grocery store. Additionally, the unpredictable nature of farming—due to factors like weather, pests, and crop diseases—can lead to variability in the quantity and quality of produce available in each share. Consumers who are not accustomed to cooking with seasonal ingredients or are unfamiliar with certain vegetables may find it difficult to utilize the produce they receive, which can lead to food waste. Some CSA programs also require upfront payment for the entire growing season, which may be prohibitive for lower-income consumers who cannot afford the initial investment, even if it ultimately saves money over time (Galt et al., 2019).

Furthermore, the infrastructure required for large-scale local food production is often lacking, particularly in urban areas. Cities, which are densely populated, have limited space for home gardens or urban farms, and access to open land for community gardens is often restricted. Establishing a home garden or CSA in urban environments may require creative solutions, such as rooftop gardens, vertical farming, or hydroponic systems, which can be complex and expensive. Additionally, urban soil may be contaminated with pollutants, posing health risks and necessitating further investment in raised beds or soil testing to ensure food safety.

Consequently, while homegrown foods provide economic and environmental benefits, practical limitations mean they may not be feasible for all consumers or food types. For many, the challenges of availability, seasonality, knowledge requirements, and resource demands create barriers to sustainable, home-based food production. As interest in local and sustainable food systems grows, addressing these challenges through educational programs, community support, and investments in infrastructure could help make homegrown foods more accessible.

However, for now, homegrown food remains a supplemental option for many rather than a comprehensive solution to food needs.

Consumer Awareness and Choices: Weighing Homegrown vs. Imported Foods

Increasing consumer awareness around the trade-offs between homegrown and imported foods empowers individuals to make informed choices that align with their environmental, health, and economic priorities. Consumers who prioritize sustainability often seek out locally grown or seasonal foods as a way to reduce their carbon footprint and contribute to regional food systems. By purchasing food grown closer to home, they support reduced food miles, which lowers the greenhouse gas (GHG) emissions associated with long-distance transportation and cold storage, particularly for highly perishable items. Locally sourced and seasonal produce can also support biodiversity and healthier soil practices, as small-scale and organic farms frequently employ sustainable agricultural methods that benefit local ecosystems (Edwards-Jones et al., 2008).

For consumers weighing the convenience and variety of imported foods, such as out-of-season fruits or exotic vegetables, understanding the environmental and economic implications can aid in making balanced choices. Imported foods offer diversity and accessibility, providing year-round access to popular items like avocados, bananas, and tropical fruits, which may not be possible to grow locally. However, this variety comes with environmental costs, as these foods require energy-intensive refrigeration, long-distance transportation, and often involve monoculture farming practices that can lead to soil degradation and higher pesticide use (McKinnon, 2018). By considering these trade-offs, consumers can make conscious decisions, such as reducing their consumption of carbon-intensive imports or purchasing them less frequently to minimize environmental impact.

Labeling systems play a critical role in guiding these choices by providing transparency around food origins and environmental impact. Origin labels, carbon footprint labels, and certifications for organic or sustainably farmed products offer consumers important information to assess the impact of their choices. For example, labels indicating food miles or CO_2-equivalent emissions can reveal the carbon footprint of imported versus locally grown foods, helping consumers opt for lower-impact options when possible. Additionally, programs that promote local food systems, such as farmers' markets, community-supported agriculture (CSA), and farm-to-table initiatives, make it easier for consumers to access fresh, local produce while supporting regional farmers. These initiatives not only reduce dependency on long-distance imports but also foster a sense of community around local food systems, encouraging sustainable consumption patterns (Weber & Matthews, 2008).

Furthermore, consumer education programs that promote understanding of seasonal eating, local sourcing, and environmentally friendly food practices can amplify the impact of informed choices. By emphasizing the benefits of eating locally and seasonally, these programs help consumers understand how their choices can influence the food industry's sustainability practices. For instance, promoting the use of local produce that is naturally in season reduces the need for energy-intensive greenhouses and artificial ripening processes, both of which contribute to GHG emissions. As consumer demand for sustainable food options grows, food retailers and producers are increasingly encouraged to adopt transparent labeling and to offer more eco-friendly choices.

By considering both environmental and economic factors, consumers can select foods that reflect their values and support sustainable practices in the food industry. Each decision, whether favoring locally sourced produce, choosing seasonal foods, or occasionally indulging in imported items, contributes to a more sustainable food system that prioritizes environmental health, local economies, and consumer awareness. With greater knowledge and access to information on food origin and carbon impact, consumers are better equipped to make choices that align with their sustainability goals and encourage positive changes within the food industry.

Chapter 9

Retail and Storage Cost of Keeping Food Fresh

R etail and storage facilities are the unsung heroes of the food supply chain, ensuring that fresh, safe, and high-quality food reaches consumers year-round. However, this crucial role comes with a significant environmental cost. Grocery stores, warehouses, and distribution centers are energy-intensive operations, requiring refrigeration, lighting, heating, ventilation, and extensive logistics to preserve food and maintain safety standards. These processes contribute substantially to greenhouse gas (GHG) emissions, with refrigeration alone accounting for nearly half of a grocery store's energy use. Additionally, the waste generated through spoiled or unsold food, along with the extensive use of packaging materials, amplifies the carbon footprint of the retail sector. As consumer demand for convenience, variety, and freshness grows, so too does the environmental impact of retail and storage facilities. This chapter explores the energy and waste dynamics of food retail, innovations that reduce carbon emissions, and the vital role of technology in creating a more sustainable and efficient food system.

Carbon Costs in Grocery Stores and Storage Facilities

Grocery stores and storage facilities play an essential role in maintaining food freshness and safety, ensuring that perishable goods reach consumers in optimal condition. However, these facilities also contribute significantly to the food industry's overall carbon footprint due to their high energy demands and operational needs. To maintain food quality, grocery stores and distribution centers consume substantial energy for refrigeration, heating, lighting, ventilation, and general operations, all of which contribute to greenhouse gas (GHG) emissions. Studies reveal that supermarkets and food storage facilities are some of the largest energy consumers within the food supply chain, with refrigeration alone responsible for up to 50% of a store's total energy usage (Heller & Keoleian, 2015). Additionally, food storage facilities, such as warehouses and distribution centers, require constant temperature regulation to prevent spoilage, which further adds to their energy consumption. As most grocery stores and storage facilities rely on electricity generated from fossil fuels, the resulting CO_2 emissions contribute to climate change, making it crucial to address the carbon costs of these operations to reduce the environmental impact of food storage and retail (Edwards et al., 2018).

In grocery stores, refrigeration systems are particularly energy-intensive and are major sources of emissions, not only because of their high electricity usage but also due to the potential leakage of refrigerants, which are extremely potent greenhouse gases. Refrigeration accounts for a significant portion of total energy consumption within retail environments, with display cases, freezers, and cold storage rooms running continuously to preserve perishable items. Traditional refrigerants, like hydrofluorocarbons (HFCs), have a global warming potential thousands of times greater than that of CO_2, meaning even small leaks can significantly increase a store's carbon footprint (McKinnon, 2018). These leaks are challenging to prevent entirely, as refrigeration systems require frequent maintenance and careful management to avoid losses. Although advancements in refrigeration technology have led to more energy-efficient and environmentally friendly systems—such as those using natural refrigerants like CO_2 or ammonia—many older systems still rely on HFCs. This reliance underscores the importance of not only advancing refrigeration technologies but also implementing strict maintenance and leak prevention practices to reduce emissions.

Beyond refrigeration, lighting and ventilation systems also contribute to the carbon footprint of retail environments. Supermarkets are typically open for extended hours, if not 24/7, requiring constant lighting to create an appealing and safe environment for customers. Traditional lighting systems, such as incandescent and fluorescent bulbs, consume significant amounts of electricity, further contributing to emissions. Modern supermarkets are increasingly adopting LED lighting, which is up to 75% more energy-efficient than traditional lighting options and emits less heat, reducing the cooling load on refrigeration systems. This shift to LED lighting allows grocery stores to cut energy consumption, decrease operational costs, and reduce their environmental impact (Smith, 2020). Ventilation and heating systems are also essential for maintaining indoor air quality and a comfortable shopping environment, though they add to the energy requirements of grocery stores. These systems are often managed through centralized control systems that optimize energy use by adjusting settings based on foot traffic, outdoor temperature, and other variables, allowing for more efficient energy use and lower emissions.

In addition to energy consumption, grocery stores and storage facilities contribute to the food industry's carbon footprint through waste generation. Food waste is particularly prevalent, as unsold perishable items are often discarded due to spoilage, cosmetic imperfections, or expiration dates. In the United States, approximately 30% of food in grocery stores is wasted, and much of this ends up in landfills, where it decomposes and releases methane—a greenhouse gas with 25 times the global warming potential of CO_2 (EPA, 2020). Many stores are taking steps to address food waste by donating surplus items to food banks, composting unsellable produce, or working with food recovery organizations to redirect food away from landfills. However, food waste remains a significant challenge, and reducing it would yield substantial carbon savings across the industry.

Packaging waste is another environmental cost associated with retail operations, as most foods are packaged to ensure hygiene and extend shelf life. This packaging, often made from plastic, cardboard, or metal, contributes to the carbon footprint of the food supply chain, both in production and disposal. Grocery stores must also manage other forms of operational waste, such as cardboard boxes, plastic wrap, and single-use materials, which require energy and resources to produce and often end up as waste. Efforts to reduce packaging waste, including bulk bins, reusable containers, and biodegradable packaging, are becoming more common, yet challenges remain in making these practices scalable and convenient for consumers.

The environmental impact of grocery stores and storage facilities underscores the importance of adopting sustainable practices and technologies to reduce energy use, improve efficiency, and minimize waste. Reducing the carbon footprint in retail environments involves a multi-faceted approach that addresses high-energy demands in refrigeration and lighting, mitigates food and packaging waste, and prioritizes leak prevention and maintenance in refrigeration systems. By investing in energy-efficient technologies, optimizing waste management, and implementing strict standards for refrigerant management, the food industry can make meaningful progress in reducing the environmental costs of retail and storage facilities.

Refrigeration, Lighting, and Waste in Retail Environments

Refrigeration and lighting are two of the most energy-intensive processes in retail environments, especially in grocery stores, where large volumes of perishable goods must be kept at stable, low temperatures to ensure safety and quality. Refrigeration in these settings includes refrigerated display cases, freezers, and cold storage rooms that run continuously, often around the clock, to maintain optimal temperatures for food preservation. This constant operation is a major contributor to a store's total electricity consumption. Beyond the direct CO_2 emissions from the electricity needed to power refrigeration systems, these units often rely on hydrofluorocarbon (HFC) refrigerants, which are potent greenhouse gases with a global warming potential thousands of times greater than CO_2. Even minor leaks of HFCs from refrigeration units can have a significant impact on a store's overall carbon footprint, underscoring the urgent need for more sustainable and reliable cooling solutions in the retail industry (Jones et al., 2017). Estimates suggest that commercial refrigeration alone accounts for roughly 1.2% of global greenhouse gas (GHG) emissions, highlighting the scale of its impact on climate change.

Lighting is another critical factor in the energy consumption and carbon footprint of retail environments. Traditional lighting systems, including incandescent and fluorescent bulbs, consume a significant amount of energy, especially in large supermarkets that operate extended hours or even around the clock. The energy demands of these lighting systems are compounded by the fact that they emit heat, which increases the cooling load on refrigeration systems, indirectly raising energy requirements for refrigeration as well. In response, many modern grocery stores have begun adopting LED lighting, which is significantly more energy-efficient than conventional lighting options. LED lighting not only consumes up to 75% less energy than traditional lighting but also has a longer lifespan, which helps reduce waste and the costs associated with frequent bulb replacements (Smith, 2020). Additionally, LED lights emit far less heat, which reduces the strain on refrigeration systems, allowing retailers to achieve indirect carbon savings by decreasing the energy required to maintain cooler ambient temperatures within the store (Energy Star, 2020). Implementing LED lighting systems across the retail sector offers substantial carbon savings, contributing to lower overall emissions from retail operations and providing an effective strategy for reducing the environmental impact of grocery stores.

Waste management is another critical element in reducing the carbon footprint of retail environments, especially in grocery stores where food waste is a prevalent issue. Unsold perishable items, particularly fruits, vegetables, and dairy products, are frequently discarded due to spoilage, expiration, or minor cosmetic imperfections. In the United States, it is estimated that approximately 30% of food in grocery stores is wasted, with much of this waste ending up in landfills where it decomposes and produces methane—a greenhouse gas that has 25 times the global warming potential of CO_2 (EPA, 2020). The environmental impact of food waste in retail settings is substantial, as methane emissions from decomposing organic matter in landfills contribute significantly to the overall carbon footprint of the food industry.

To address the issue of food waste, many retailers are implementing food waste reduction programs that aim to divert unsold food from landfills and mitigate methane emissions. These initiatives often include partnerships with food banks and other charitable organizations, which allow unsellable but still edible food to be donated to those in need. Some grocery stores are also implementing composting programs to repurpose food waste into valuable organic material that can enrich soil and support sustainable agriculture. Additionally, many retailers are exploring partnerships with food recovery organizations, which specialize in redistributing surplus food to communities and individuals who need it, further reducing the volume of waste sent to landfills. By adopting these strategies, retailers not only reduce their carbon footprint but also contribute to food security and waste reduction efforts, creating a positive impact on both the environment and society.

The combined impact of sustainable refrigeration, energy-efficient lighting, and effective waste management strategies can significantly reduce the carbon footprint of grocery stores and retail environments. As the industry adopts more sustainable practices and technologies, the potential for meaningful emissions reductions becomes more attainable, aligning retail operations with broader environmental goals and contributing to a more sustainable food supply chain.

The Role of Technology in Reducing Carbon Output in Retail

Advances in technology are enabling grocery stores and storage facilities to significantly reduce their carbon footprint through enhanced energy efficiency, improved waste management, and the adoption of sustainable practices. These innovations help address key areas of energy consumption and emissions within the food retail industry, transforming traditional processes to align with environmental goals and reduce operational costs.

Refrigeration Innovations: Eco-Friendly Refrigerants and Smart Systems

Refrigeration technology has been one of the primary focuses for reducing carbon emissions, as it represents a major energy drain in grocery stores and storage facilities. Traditional refrigerants, particularly hydrofluorocarbons (HFCs), are powerful greenhouse gases with high global warming potentials, exacerbating the environmental impact of refrigeration systems. One key area of innovation involves replacing HFCs with more eco-friendly refrigerants, such as CO_2 and ammonia, which have a much lower global warming potential. These natural refrigerants are not only more environmentally friendly but also allow cooling systems to operate at high efficiency with lower emissions (McKinnon, 2018). Some grocery chains have started retrofitting older systems to accommodate these refrigerants or installing new systems that rely on them, aligning refrigeration practices with sustainability targets.

In addition to more sustainable refrigerants, advancements in smart refrigeration systems equipped with sensors, Internet of Things (IoT) connectivity, and automated controls are enhancing energy efficiency in cooling processes. These systems allow for precise temperature regulation by adjusting cooling needs based on real-time conditions, such as fluctuations in store traffic or ambient temperatures. By maintaining optimal temperatures only when necessary, smart refrigeration can significantly reduce energy consumption. Additionally, IoT-enabled systems monitor for leaks in real time and optimize compressor performance, reducing the risk of refrigerant loss and ensuring that equipment operates at peak efficiency. This proactive monitoring helps prevent the unintentional release of refrigerants, further decreasing greenhouse gas emissions (Jones et al., 2017). Overall, these refrigeration technologies represent a significant shift toward energy-efficient cooling solutions in retail environments.

Energy Management Systems (EMS): Reducing Carbon Output Across Operations

Energy management systems (EMS) are another critical technological advancement helping grocery stores and storage facilities reduce their carbon footprint. EMS solutions enable retailers to monitor, track, and control energy usage across multiple processes, including refrigeration, lighting, heating, and ventilation. By collecting and analyzing energy data, EMS can identify inefficiencies, alert managers to areas of high energy usage, and implement targeted reductions. For example, EMS can automatically dim lighting during low-traffic periods or adjust HVAC systems based on real-time weather conditions, thus conserving energy during times when full power is unnecessary. These systems can even integrate with renewable energy sources, such as solar panels, allowing stores to optimize energy use based on the availability of clean energy. Many large grocery chains have adopted EMS to successfully reduce their overall energy consumption by as much as 20%, underscoring the effectiveness of these systems in lowering emissions and operational costs (Smith, 2020).

Digital Inventory Management: Cutting Down on Food Waste

Beyond energy use, technology is also playing a vital role in waste reduction within retail environments. Digital inventory management systems use data analytics and machine learning to accurately track product lifespans, forecast demand, and optimize ordering schedules, which helps reduce overstocking—a primary contributor to food waste. By predicting sales patterns more precisely and managing stock in real time, these systems ensure that products are ordered in quantities that match demand, thus minimizing the risk of items reaching expiration before they are sold. Inventory management systems are often integrated with point-of-sale (POS) data, providing managers with a comprehensive view of inventory levels, turnover rates, and upcoming expiration dates.

In addition, automated markdown systems help further reduce waste by adjusting prices for products nearing their sell-by dates. This real-time price adjustment incentivizes consumers to purchase soon-to-expire items at discounted rates, which increases turnover and decreases the volume of food that might otherwise be discarded. Some retailers are even experimenting with "smart shelves" that notify managers when perishable items are nearing expiration, allowing for timely price reductions or promotions. By reducing food waste, inventory management technologies not only save resources but also cut down on the carbon emissions associated with food disposal, including methane emissions from food that would otherwise decompose in landfills (EPA, 2020).

Adopting Renewable Energy and Sustainable Building Practices

In addition to energy efficiency and waste management, some grocery stores and storage facilities are adopting renewable energy sources and sustainable building practices to further reduce their environmental impact. Solar panels, for instance, are increasingly being installed on rooftops of large grocery stores, helping to power energy-intensive operations with clean energy. In locations with high solar exposure, solar power can contribute a substantial portion of the store's electricity needs, reducing dependence on fossil fuels. Energy-efficient building materials, enhanced insulation, and heat recovery systems are also becoming more common, helping to improve energy retention and decrease heating and cooling costs in large retail spaces (Energy Star, 2020). These sustainable building practices not only lower energy requirements but also contribute to a more resilient infrastructure capable of withstanding temperature fluctuations and reducing the overall environmental impact.

In sum, technology is driving significant advancements in reducing carbon emissions in retail environments, addressing various facets of store operations from refrigeration and lighting to inventory and waste management. The adoption of eco-friendly refrigerants, smart refrigeration systems, energy management systems, digital inventory management, and renewable energy are transforming grocery stores and storage facilities, enabling them to align with broader sustainability goals. As these innovations become more widely adopted across the food industry, the potential to make meaningful progress in lowering the sector's carbon footprint is within reach. Through technology-driven efficiency and a shift toward sustainable practices, the food retail industry is evolving toward a more climate-conscious and resource-efficient future, contributing to a more sustainable food supply chain.

Chapter 10

Consumer Cost of Choice and Waste

Consumer behavior is one of the most influential factors shaping the modern food system. From dietary preferences and purchasing habits to the way food is stored, prepared, and discarded, individual choices ripple through every stage of the food supply chain. These preferences dictate not only what is grown but also how it is processed, packaged, and transported, directly impacting the environmental footprint of the global food industry. However, this power comes with significant responsibility. The demand for variety, convenience, and abundance has driven unsustainable practices, including the overproduction of resource-intensive foods, excessive packaging waste, and high rates of food spoilage. As a result, consumers play a pivotal role in addressing some of the most pressing challenges in the food system, such as reducing greenhouse gas emissions, minimizing food waste, and supporting sustainable agriculture. This chapter delves into how consumer preferences shape the food chain, the economic and carbon costs of food waste, and practical strategies for adopting more sustainable shopping and eating practices to align individual choices with broader environmental goals.

How Consumer Preferences Shape the Food Supply Chain

Consumer preferences have a profound influence on the structure and environmental impact of the food supply chain, dictating not only what types of foods are grown and how they are processed but also the methods used to transport, package, and display them in retail settings. With an increasing demand for year-round availability of diverse fruits, vegetables, and fresh products, food suppliers are often compelled to source these items from regions where they can be produced regardless of the season. This global demand for variety and convenience adds significant complexity to the supply chain, as foods such as avocados, berries, and tropical fruits often travel thousands of miles from countries like Mexico, South America, and Southeast Asia. Transporting perishable goods across long distances typically requires high-energy refrigeration and, in some cases, air freight—an especially carbon-intensive mode of transportation. This reliance on long-distance importation to meet consumer demand results in a high carbon footprint for these products, contributing substantially to greenhouse gas (GHG) emissions in the food sector (Heller & Keoleian, 2015).

The consumer preference for visually perfect produce also drives inefficiencies and waste within the supply chain. Supermarkets often adhere to stringent aesthetic standards to meet consumer expectations, resulting in the discarding of blemished, misshapen, or slightly damaged fruits and vegetables, which are perceived as unappealing. As a consequence, a considerable portion of produce is discarded before it even reaches the retail floor, wasting resources and contributing to GHG emissions from both the growing and disposal stages. This standardization for visual perfection means that otherwise edible and nutritious food is lost, adding to the environmental cost of agricultural production, as water, fertilizer, and labor are expended on crops that are ultimately discarded (Garnett, 2011). Such wasteful practices underscore how consumer expectations around appearance indirectly drive food waste and carbon emissions along the entire supply chain.

Packaging is another significant area of influence shaped by consumer expectations for convenience and product longevity. In response to consumer demand for easy-to-use, durable packaging, retailers frequently employ single-use plastics and other non-biodegradable materials. While plastic packaging is effective at preserving freshness and extending shelf life, it also contributes to the growing problem of plastic pollution and landfill waste. This issue is particularly prevalent in countries where recycling systems are underdeveloped, leading to high rates of plastic entering natural ecosystems. Plastic takes hundreds of years to break down, often fragmenting into microplastics that contaminate soil, waterways, and marine environments. Consumer reliance on convenience and the increasing popularity of single-use, pre-packaged, and ready-to-eat meals drive up the amount of packaging required per unit of food, further exacerbating the environmental footprint of the food industry (Robertson, 2012).

The growing popularity of pre-packaged, individually wrapped foods highlights a shift in consumer behavior toward convenience, often at the expense of increased waste and carbon emissions. This trend is particularly evident in the rising demand for "grab-and-go" options and portion-controlled packages, which cater to busy lifestyles but require excessive packaging relative to the amount of food provided. For example, pre-sliced fruits in plastic containers, single-serving snack packs, and meal kits all contribute to an increased reliance on packaging, creating a cycle of waste as these single-use items are quickly discarded. Additionally, the resources and energy needed to produce and transport packaging materials like plastic, glass, and aluminum are substantial, contributing to the carbon footprint of each product beyond the food itself. As these convenience-driven choices become more prevalent, they reinforce practices within the food supply chain that prioritize accessibility and ease over environmental sustainability (Garnett, 2011).

In addition to packaging and transportation, consumer demand for a consistent and abundant selection of foods has led to greater reliance on monoculture farming, which can lead to soil depletion, reduced biodiversity, and increased use of chemical fertilizers and pesticides. Monoculture farming—growing a single crop extensively across large tracts of land—is often economically efficient but can degrade soil health and require synthetic inputs to maintain yields, further contributing to carbon emissions and environmental harm. The desire for crops like wheat, corn, and soybeans, which are used in a wide range of processed foods, has resulted in farming practices that prioritize high yields and uniformity, aligning with consumer expectations but often at the expense of ecological sustainability (Tilman et al., 2011). This demand for high-output farming practices demonstrates how consumer choices can influence agricultural methods that have far-reaching impacts on soil health, water resources, and GHG emissions.

Ultimately, consumer preferences shape food supply chains that prioritize availability, aesthetic appeal, and convenience—often leading to environmental trade-offs such as increased waste, pollution, and carbon emissions. Raising awareness about the impact of these choices and shifting consumer attitudes toward more sustainable options, such as seasonal and local foods, minimal packaging, and reducing cosmetic standards, could help reshape the food supply chain in a way that supports environmental sustainability. Educating consumers about the true cost of their choices may inspire a shift toward practices that reduce waste, lower carbon footprints, and contribute to a more resilient food system.

Food Waste at Home: Economic and Carbon Implications

Food waste at home remains one of the most significant challenges to reducing the environmental impact of food consumption, particularly in developed nations where households contribute substantially to the overall waste stream. Factors such as over-purchasing, improper storage, and a lack of meal planning often lead to food being discarded before it is consumed. In the United States, for instance, studies show that between 30% and 40% of food in households is wasted, which translates to both economic and environmental losses (Buzby & Hyman, 2012). Economically, reducing food waste could save households hundreds of dollars per year, as studies estimate that the average American household wastes food worth hundreds, if not thousands, of dollars annually. These savings could be realized through simple changes such as shopping with a list, storing food properly, and using leftovers creatively to minimize food loss.

The carbon implications of food waste are even more severe. Food waste represents a loss of not only the food itself but also all the resources—water, energy, and labor—that went into its production, processing, and transport. This "hidden" carbon cost becomes especially apparent when food waste is disposed of in landfills, where it decomposes anaerobically (without oxygen) and releases methane, a greenhouse gas with a global warming potential 25 times greater than CO_2 (EPA, 2020). Food waste accounts for a significant share of methane emissions from landfills, which are one of the largest sources of this potent greenhouse gas. Reducing food waste at home could therefore play a critical role in decreasing methane emissions, offering a straightforward way for households to contribute to climate change mitigation.

Furthermore, food waste at home undermines sustainability efforts made across the entire food supply chain. While efforts at the production, transportation, and retail stages may focus on reducing emissions and improving efficiency, household food waste negates these advances, as all the embedded resources and emissions in the discarded food are lost. For example, a single avocado wasted at home represents not only the emissions from its transport but also the water, fertilizer, and labor required to grow it, as well as the energy and materials used in packaging and refrigeration. This cumulative impact of wasted resources makes household food waste a significant bottleneck in the sustainability of the food system (Garnett, 2011). Addressing this challenge at the consumer level is therefore crucial to achieving a more sustainable food system, as reductions in household food waste can directly lead to lower emissions and resource conservation.

Efforts to reduce food waste at home have the potential to generate environmental, economic, and social benefits. Initiatives such as educating consumers about proper food storage techniques, providing tools for meal planning, and promoting the use of leftovers could significantly reduce household food waste. Additionally, awareness campaigns that highlight the hidden costs of food waste—both financial and environmental—can encourage consumers to adopt more sustainable behaviors. Implementing community-level programs, such as composting and food-sharing networks, can further support these efforts, providing consumers with practical solutions for diverting food waste from landfills and making better use of surplus food.

Addressing food waste at home offers a direct and impactful way to improve the sustainability of the food system, conserve valuable resources, and reduce greenhouse gas emissions. By taking responsibility for food choices and adopting waste-reducing habits, consumers can play an essential role in mitigating the environmental impact of food waste and promoting a more sustainable approach to food consumption.

Sustainable Shopping and Eating Practices

Sustainable shopping and eating practices offer consumers practical strategies to reduce their environmental impact while aligning their food choices with broader sustainability goals. By adopting mindful approaches to food purchasing, meal planning, and consumption, individuals can make a positive impact on the food system and minimize waste, pollution, and emissions associated with food production and disposal.

Meal Planning and Conscious Purchasing

One of the most effective changes consumers can make to reduce their environmental impact is to plan meals ahead of time and shop with a list, which helps minimize the risk of over-purchasing and reduces food spoilage. When consumers buy only what they need, they lower household food waste, decrease the demand for excessive food production, and save money. Meal planning also encourages consumers to utilize leftovers creatively, reducing waste further by repurposing extra ingredients. Additionally, shoppers who make thoughtful purchasing decisions are more likely to avoid impulse buys and unnecessary packaged items, both of which contribute to waste (Buzby & Hyman, 2012). Meal planning thus creates a cycle of mindful consumption that benefits both the environment and household finances.

Seasonal and Locally Sourced Produce

Choosing seasonal and locally sourced produce is another impactful way to reduce the carbon footprint associated with food consumption. Local, seasonal foods are typically grown under natural climate conditions and harvested at peak freshness, which reduces the need for artificial inputs like greenhouses, fertilizers, and pesticides. Seasonal produce also requires shorter transportation distances, thereby decreasing fuel consumption and emissions associated with long-haul freight (Weber & Matthews, 2008). When consumers buy locally, they support regional farmers and foster a connection to their community, helping to strengthen local food systems and reduce the dependency on large-scale, resource-intensive agricultural practices. Furthermore, because seasonal foods are often fresher and have a shorter shelf life, they tend to retain more nutritional value, which can encourage healthier eating.

Adopting Plant-Based or Flexitarian Diets

Consumers can further support sustainable food systems by adopting plant-based or flexitarian diets, both of which emphasize plant foods while reducing or moderating animal product consumption. Plant-based foods have a significantly lower carbon footprint than animal products, as they generally require fewer resources, including land, water, and energy, for production. Shifting towards a diet rich in fruits, vegetables, legumes, and grains can greatly reduce individual carbon footprints, as well as the demand for environmentally taxing animal agriculture. Studies suggest that reducing meat and dairy consumption—even without fully eliminating it—can lead to substantial environmental benefits (Heller & Keoleian, 2015). Flexitarian diets, which include moderate amounts of animal products, offer an accessible alternative that balances environmental impact with dietary preferences, allowing consumers to make meaningful changes without drastic lifestyle shifts.

For those who continue to consume animal products, choosing sustainably raised options, such as pasture-raised or organically farmed meat, dairy, and eggs, can help lower environmental impact. These methods often involve better land management practices, like rotational grazing and reduced reliance on synthetic fertilizers and pesticides, which promote soil health, conserve water, and reduce pollution. Supporting producers who prioritize animal welfare and sustainable practices not only minimizes ecological harm but also encourages more responsible agricultural methods in the food industry.

Mindful Consumption and Waste Reduction Practices

Mindful consumption practices, such as using reusable bags, buying unpackaged goods, and selecting items with minimal or recyclable packaging, are essential to reducing waste in the food system. By choosing bulk items or products in reusable or biodegradable containers, consumers can significantly cut down on plastic and single-use packaging waste, supporting a zero-waste lifestyle that aligns with the goals of sustainable consumption. Reusable bags, containers, and produce bags help reduce plastic pollution, which is a major environmental issue with impacts on marine life, soil health, and ecosystems. Furthermore, buying in bulk reduces the overall demand for packaging materials, aligning with principles of waste reduction and resource efficiency (Smith, 2020).

Many consumers are also participating in community-supported agriculture (CSA) programs or shopping at farmers' markets, which provide access to fresh, local produce directly from farmers. CSAs offer seasonal produce that is often grown using sustainable practices, reducing food miles and supporting the economic viability of local farms. Farmers' markets also provide an alternative to conventional retail settings, fostering direct connections between consumers and food producers. This support for local farmers not only strengthens community ties but also reduces reliance on long-distance transportation and large-scale agriculture, creating a more resilient and sustainable food system.

Supporting Food Recovery and Food Sharing Initiatives

In addition to making sustainable shopping choices, consumers can support food recovery initiatives to reduce food waste at a community level. Food recovery programs, such as food banks and food-sharing networks, collect surplus food from retailers, restaurants, and households, redistributing it to people in need instead of allowing it to go to waste. By participating in or donating to these programs, consumers contribute to reducing food waste and its associated environmental impacts, including methane emissions from decomposing food in landfills. Food sharing platforms, such as mobile apps that connect individuals with excess food to nearby recipients, are also gaining popularity as a way to reduce household food waste while fostering community connections (EPA, 2020).

By adopting sustainable shopping and eating practices, consumers can make choices that support a resilient, low-impact food system. Meal planning, choosing seasonal and local produce, embracing plant-based diets, reducing packaging waste, and supporting food recovery initiatives all contribute to reducing personal and collective carbon footprints. As consumers become more aware of their impact on the environment, these mindful practices offer a pathway to meaningful change, fostering a food system that prioritizes sustainability, waste reduction, and community support. Sustainable consumer behavior has the potential to reshape the food industry, making it more responsive to the environmental challenges of our time and promoting a healthier, more equitable food system for future generations.

Chapter 11

End of the Food Life Cycle Waste Management

T he food system's environmental and economic footprint does not end at the dinner table; rather, it culminates in the disposal of uneaten food and the packaging left behind. The issue of food waste, particularly in developed nations, represents one of the most glaring inefficiencies in the global food chain, contributing significantly to greenhouse gas (GHG) emissions and resource depletion. From methane emissions produced in landfills to the wastage of water, energy, and labor embedded in discarded food, the costs of waste are both staggering and preventable. This chapter examines the role of food waste in contributing to carbon emissions, explores innovative alternatives such as composting and anaerobic digestion, and highlights the critical role of policies in mitigating waste. By addressing waste at the end of the food life cycle, we can unlock significant environmental and economic benefits, fostering a system that respects resources and minimizes its impact on the planet.

Food Waste and Its Impact on Carbon Emissions

Food waste stands out as one of the largest contributors to greenhouse gas (GHG) emissions within the food life cycle, with repercussions that extend across both environmental and economic domains. When food waste is discarded in landfills, it decomposes anaerobically—meaning in the absence of oxygen—a process that leads to the release of methane. Methane is a particularly potent greenhouse gas, with a global warming potential that is up to 25 times greater than that of carbon dioxide over a 100-year period (EPA, 2020). In the United States alone, food waste contributes to roughly 8% of total GHG emissions, largely due to methane emissions from landfills. This significant impact places food waste among the top contributors to landfill methane emissions, underscoring the urgent need for measures to divert organic waste away from landfills and thus reduce its role in climate change (Heller & Keoleian, 2015).

Beyond the release of methane, food waste represents a colossal waste of resources that were originally invested in food production. When edible food is discarded, so too are the resources—such as water, energy, fertilizer, and labor—that were required to cultivate, harvest, process, and transport it. For instance, the production of a single kilogram of beef requires approximately 15,000 liters of water, representing a tremendous resource investment. When that beef is wasted, all of that water, along with the energy for processing and transportation, is also wasted (Buzby & Hyman, 2012). This inefficiency is not limited to animal products. Fruits and vegetables, which require substantial inputs of water, fertilizer, and labor, similarly squander these resources when they end up in the trash.

The impact of food waste extends to energy resources as well. A substantial amount of energy goes into processing, packaging, refrigerating, and transporting food items, especially perishable goods like dairy, meat, and fresh produce. When these foods are discarded, the energy expended across each stage of the supply chain—from farming to the supermarket—is effectively wasted. This inefficiency amplifies the carbon footprint of food production, as it means that the fossil fuels burned to produce and deliver that food ultimately contribute to emissions without fulfilling the food's intended purpose (Garnett, 2011).

The scale of resource waste is particularly notable in industrialized nations, where food waste tends to occur primarily at the consumer level. In contrast, in developing countries, food loss is more common during post-harvest handling and storage due to inadequate infrastructure. This distinction highlights the need for tailored strategies to reduce food waste according to the context of each region. For example, in high-income countries, where a significant portion of food waste happens at the consumer level, education on meal planning, proper food storage, and portion control can play a pivotal role in minimizing waste. In low-income countries, however, investment in improved storage, transportation, and preservation technologies is essential to prevent food loss before it reaches the consumer (Gustavsson et al., 2011).

Additionally, the financial implications of food waste are significant. On a household level, reducing food waste could save individuals hundreds of dollars per year. At a societal level, the economic burden of food waste includes the costs associated with its disposal and the resources invested in production that yield no return when food is wasted. The Food and Agriculture Organization of the United Nations estimates that global food waste accounts for approximately $750 billion in economic losses annually, a figure that underscores the broader impacts of waste on economies worldwide (FAO, 2013). By addressing food waste, we not only conserve environmental resources but also create opportunities for economic efficiency, particularly within industries that manage food production, processing, and distribution.

Reducing food waste is therefore essential not only for decreasing emissions and resource use but also for building a more resilient food system that can sustainably meet global food demands. A multifaceted approach—encompassing waste diversion strategies, improved food management practices, and sustainable consumer behaviors—can significantly mitigate the impact of food waste on the environment. Strategies such as composting, food recovery, and investment in waste prevention infrastructure are critical to redirecting food waste from landfills, thereby reducing emissions and promoting a circular economy in which resources are used efficiently and effectively across their life cycle.

Composting, Recycling, and Landfill Alternatives

Composting and recycling food waste provide effective alternatives to traditional landfill disposal, transforming food waste into valuable resources rather than allowing it to release methane emissions that contribute to climate change. Composting is a natural decomposition process that uses oxygen to break down organic materials, such as food scraps and yard waste, into nutrient-rich soil amendments. Unlike anaerobic decomposition in landfills, which occurs without oxygen and results in methane production, composting relies on aerobic conditions that prevent methane emissions. By returning essential nutrients to the soil, composting enhances soil health, improving its structure, moisture retention, and nutrient content, which are vital for sustainable agriculture and plant growth (EPA, 2020). Composting retains organic materials that would otherwise contribute to GHG emissions, and it supports soil ecosystems by increasing biodiversity and encouraging beneficial microorganisms.

Composting can be implemented at multiple levels, making it accessible for both individuals and larger communities. Home composting, for instance, allows individuals to divert food waste from landfills, providing a direct way for households to reduce their environmental impact. Many municipalities offer community composting programs, often in collaboration with local gardens or farms, where residents can drop off food scraps to be processed into compost for public use. Larger-scale composting facilities, known as industrial composting, process high volumes of organic waste, including food from restaurants, grocery stores, and agricultural sources. Industrial composting can handle a wider range of materials, such as meat and dairy, which are not typically suited for home composting, making it a more versatile option for organic waste recycling (Gunders et al., 2017).

Some municipalities and regions are taking composting a step further by implementing food waste recycling programs like anaerobic digestion (AD). AD is a process that uses specific microorganisms to break down organic materials in a controlled, oxygen-free environment, capturing methane as it is produced. This methane is then converted into biogas, which can be used as a renewable energy source to replace fossil fuels in power generation, heating, or even as vehicle fuel. Biogas production through anaerobic digestion not only reduces GHG emissions but also recycles waste into a valuable energy source, supporting a circular economy that recovers and reuses resources. The byproduct of anaerobic digestion, known as digestate, is rich in nutrients and can be used as a high-quality organic fertilizer in agriculture, further enhancing the circular use of resources (Zaman & Lehmann, 2013). This process benefits both energy and agricultural sectors, creating renewable energy while enriching soil health.

In addition to composting and AD, other landfill alternatives include waste-to-energy (WTE) facilities, which incinerate organic waste to produce electricity. WTE technology has been used to reduce the volume of waste that goes to landfills while recovering energy from materials that would otherwise be discarded. In WTE plants, organic waste is burned at high temperatures to produce heat, which is then used to generate steam and electricity. This incineration process reduces the waste volume significantly, helping to alleviate the strain on landfills and generate energy. However, WTE facilities are often controversial due to the pollutants that can be released during incineration, including carbon dioxide, nitrogen oxides, and particulate matter. Modern WTE plants incorporate advanced filtration and emission control technologies to reduce harmful outputs, but they still face criticism from environmental advocates who argue that these facilities may discourage waste reduction and recycling efforts by creating a market for waste as fuel (Hodge, 2017).

Despite the energy recovery potential of WTE, composting and anaerobic digestion are generally considered more sustainable options for managing food waste. Composting and AD maximize resource recovery by returning nutrients to the soil and producing renewable energy with fewer emissions than incineration. These processes align with principles of sustainability by promoting a circular economy in which resources are continually reused and reintegrated into the natural environment, reducing reliance on landfills and minimizing pollution. Through local composting initiatives, municipal AD facilities, and community-based programs, cities and regions can improve their waste management practices, support local agriculture, and reduce GHG emissions.

By integrating composting, anaerobic digestion, and waste-to-energy solutions, communities can implement a comprehensive approach to food waste management. Such systems not only divert organic materials from landfills but also contribute to environmental, economic, and social benefits, such as renewable energy production, improved soil quality, and reduced carbon emissions. These landfill alternatives reflect a shift toward sustainable waste management that prioritizes resource recovery and environmental protection, positioning food waste as an opportunity for ecological and economic growth rather than as mere refuse.

The Role of Policies in Reducing Food Waste

Policy initiatives are essential in the battle to manage food waste, promote waste reduction, and create a sustainable food system. At multiple levels of government, from local municipalities to national programs, policies are being implemented to address the pressing environmental and social issues associated with food waste. These initiatives are often structured around incentives for businesses, regulations that restrict certain wasteful practices, and public awareness campaigns that aim to shift behaviors toward more sustainable practices.

For instance, in the United States, the Environmental Protection Agency (EPA) and the Department of Agriculture (USDA) have set ambitious targets through the U.S. Food Loss and Waste 2030 Champion program, aiming to cut food waste in half by 2030. This program brings together businesses, organizations, and governmental agencies to commit to reducing waste, promoting efficient food use, and adopting sustainable waste management practices. By encouraging businesses to prioritize waste prevention, the program sets an example for industries that might otherwise overlook waste reduction due to economic or logistical constraints. Many businesses in the food supply chain—such as supermarkets, food processors, and restaurants—have signed on as partners, committing to measures that reduce their waste footprint through improvements in storage, inventory management, and food donation initiatives (EPA, 2020).

Landfill Restrictions and Zero-Waste Policies

A growing number of cities and countries are adopting policies to restrict organic waste disposal in landfills, supporting a shift toward zero-waste goals. In San Francisco, for example, a pioneering Zero Waste Program mandates that businesses and households separate their waste into recycling, compost, and landfill categories, with penalties for non-compliance. The program has been remarkably successful in diverting waste from landfills and increasing composting rates, setting an example for other cities worldwide. The mandatory composting initiative not only decreases the amount of food waste going to landfills but also supplies nutrient-rich compost for use in local agriculture, landscaping, and gardening, closing the loop in the food waste cycle (Liptak, 2019). Other cities, such as Seattle and New York, have adopted similar policies, incentivizing composting while discouraging landfill disposal through fines and increased public access to composting facilities.

France is leading the way in waste prevention with progressive legislation that requires large supermarkets to donate unsold but edible food to charities. Passed in 2016, this law prevents large retail stores from discarding edible food and imposes fines for non-compliance. Instead of sending food to waste, supermarkets are legally bound to donate to food banks and other charities, directly addressing both food waste and food insecurity (Mourad, 2016). This policy has proven effective, reducing the amount of food waste generated by retailers while helping to feed those in need. Several other European countries, such as Italy and Spain, have enacted similar laws, encouraging food recovery practices that benefit local communities while conserving resources. These policies reflect a growing recognition of the interconnected nature of food waste and food insecurity and the potential for policy to create a more socially responsible food system.

Education and Awareness Programs

Education and public awareness are central to effective food waste reduction policies, as they address behaviors and cultural attitudes that contribute to waste. Many policies now incorporate educational campaigns to inform consumers and businesses about the environmental and economic impacts of food waste, as well as strategies for reducing waste in daily life. For example, campaigns might focus on teaching consumers about proper food storage techniques, portion control, and ways to repurpose leftovers. Some school programs are introducing food waste education into curricula to foster sustainable habits in children from a young age. By building awareness, these programs aim to shift social norms around food waste and encourage individuals to view food as a valuable resource rather than a disposable commodity.

Public awareness campaigns also encourage businesses to adopt waste-reducing practices. Programs that promote "ugly" or imperfect produce, for instance, help challenge the stigma around visually imperfect foods, encouraging retailers and consumers to buy and use food that would otherwise be discarded. Supermarkets and grocery stores can also benefit from consumer education campaigns, as they may increase demand for less cosmetically appealing but equally nutritious food options, reducing food waste at the retail level (FAO, 2013). Some governments also support research into innovative technologies for waste recycling and recovery, further enhancing policy effectiveness by backing developments in waste-to-energy conversion, anaerobic digestion, and more efficient composting processes.

Supporting Innovation and Research

To address the technical and logistical challenges of food waste management, some governments are investing in research and development (R&D) initiatives that explore new methods for recycling, reusing, and reducing food waste. This support for innovation is particularly important in urban areas, where space limitations and high waste volumes can make traditional composting or recycling difficult. Government funding for R&D has led to advancements in anaerobic digestion technology, enhanced food recovery systems, and improved bioplastics for packaging that reduces spoilage without contributing to plastic pollution. By supporting sustainable waste technologies, policy initiatives can help scale solutions that reduce the environmental footprint of food waste (Zaman & Lehmann, 2013).

Some policies also provide incentives and subsidies to companies that adopt waste-reducing technologies or sustainable practices. For instance, certain tax incentives or grants are available to companies that donate food or implement sustainable packaging solutions. By making it financially attractive for businesses to participate in waste reduction, these policies can shift industry practices and encourage companies to view sustainability as part of their operational framework rather than a costly add-on. Incentive-based policies foster a proactive approach within the food industry, helping to establish waste prevention as a standard practice.

Global and National Collaborations on Food Waste

At the international level, collaborations among countries are proving instrumental in tackling food waste. The United Nations, through its Sustainable Development Goals (SDGs), has prioritized food waste reduction, aiming to cut global food waste per capita in half by 2030 (UNEP, 2020). This initiative encourages nations to set food waste reduction targets and adopt practices that align with sustainable development. International partnerships provide platforms for sharing knowledge, research, and successful policy strategies, helping nations address food waste with a united, global approach.

In summary, policies that address food waste at local, national, and international levels are essential for achieving meaningful reductions in waste and emissions. By promoting practices such as composting, food recovery, landfill restrictions, and educational initiatives, these policies drive systemic changes that reduce the environmental impact of food waste. As more regions adopt and adapt policies to curb food waste, the cumulative effect is expected to bring about a more sustainable, resource-efficient, and equitable food system.

Chapter 12

Comparing Local and Global Food Chains

The interconnectedness of our global food system brings both remarkable diversity to our diets and significant challenges to the planet. As food travels thousands of miles from farms to tables, the environmental and economic impacts of these journeys accumulate, shaping the carbon footprint of every meal. Local food systems, on the other hand, offer an alternative by prioritizing shorter supply chains that reduce emissions and support regional economies. Understanding the carbon and cost differences between locally sourced and imported foods, along with the opportunities presented by innovative models such as community-supported agriculture (CSA) and urban farming, allows us to critically evaluate the trade-offs and benefits of each approach. By addressing the challenges of localizing food systems and integrating solutions, we can envision a future that balances sustainability, accessibility, and economic growth.

The Carbon and Cost Comparisons Between Locally Sourced and Imported Foods

When evaluating the environmental and economic impacts of food systems, comparing the carbon footprints and costs of locally sourced versus imported foods provides crucial insights into how food sourcing choices can shape both sustainability and affordability. Locally sourced foods typically have a smaller carbon footprint due to shorter transportation distances, which directly reduce "food miles" and the associated greenhouse gas (GHG) emissions. Food miles—a measure of the distance food travels from its point of production to the consumer—strongly influence the carbon impact of food items. Studies consistently show that transportation can account for a significant share of the emissions associated with imported foods, especially for perishable goods requiring energy-intensive preservation methods, such as refrigeration or climate-controlled storage. These goods are often transported by air freight, which is particularly carbon-intensive compared to other methods of transport, such as shipping or trucking. As an example, tropical fruits or off-season vegetables imported from distant regions may generate a carbon footprint up to five times greater than that of locally grown alternatives due to the energy and fuel demands needed to preserve freshness during long-haul transport (Weber & Matthews, 2008).

Beyond transportation-related carbon emissions, the economic cost of imported foods is also often higher and subject to volatility due to several factors, including fuel prices, international trade policies, and tariffs. These fluctuations can make imported foods less predictable in price, impacting consumer budgets and potentially limiting access to imported items when prices spike. Imported goods frequently incur higher costs due to the need for long-haul transportation, refrigerated storage, and specialized packaging, all of which add to the product's final price at the store. Additionally, the complexity of global supply chains—often involving multiple intermediaries, warehouses, and logistical arrangements—introduces more variables that can drive up costs, making imported foods more expensive and vulnerable to supply disruptions.

In contrast, locally sourced foods often require fewer intermediaries and, thus, can offer lower transportation costs, especially in regions with well-developed local food networks. Shorter supply chains reduce reliance on complex logistical networks, allowing producers to deliver directly to

101

consumers or through community-centered channels. For example, community-supported agriculture (CSA) programs, farmers' markets, and direct-to-consumer models are examples of localized distribution systems that help minimize both emissions and costs. CSA programs, where consumers buy shares of a farm's harvest and receive regular deliveries of fresh produce, offer economic benefits by providing farmers with upfront financial support while reducing transportation emissions. Similarly, farmers' markets and local food co-ops allow producers to sell directly to consumers, cutting out intermediaries and minimizing the need for extensive packaging or refrigeration (Coleman-Jensen et al., 2019).

However, while locally sourced foods often reduce carbon emissions and, in some cases, costs, they may not always be the most economical option. Local production can be associated with higher labor and land costs, particularly in regions where wages and land values are high. For instance, urban or peri-urban farms may face elevated costs for labor and leasing land compared to larger rural farms that benefit from economies of scale and lower real estate prices. These cost factors can make some locally produced items more expensive than imported counterparts, highlighting the complex balance between economic and environmental factors in food sourcing. For consumers, this complexity means that the benefits of local food must be weighed against higher prices, availability, and seasonal limitations. Local foods are generally more accessible during peak growing seasons and may become scarcer and more expensive during off-seasons, a limitation that does not affect imported items, which can be sourced from regions with year-round growing conditions (Garnett, 2011).

The environmental and economic differences between local and global food sourcing underscore the importance of building resilient food systems that can incorporate both local and global networks. While local food systems can significantly reduce carbon emissions, support regional economies, and offer fresher produce with minimal processing, global food chains provide access to a broader range of food items and support economic stability through international trade. Policymakers and consumers alike are beginning to recognize the potential of hybrid food systems that prioritize local production when possible but complement it with responsible imports that consider both economic and environmental impacts.

Case Studies of Community-Supported Agriculture and Urban Farming

Community-supported agriculture (CSA) and urban farming initiatives provide effective and innovative examples of localized food systems that aim to mitigate the environmental impacts of food production, enhance food security, and stimulate local economies. These models represent a shift from large-scale, industrialized food production toward systems that prioritize local, sustainable, and community-centered approaches.

Community-Supported Agriculture (CSA)

Community-supported agriculture (CSA) is a subscription-based farming model that enables consumers to directly invest in a local farm by purchasing shares of the upcoming harvest. In exchange, shareholders receive regular boxes of fresh, seasonal produce throughout the growing season. This arrangement offers substantial benefits for both farmers and consumers. For farmers, CSAs provide much-needed upfront capital to cover production costs, including seeds, labor, and equipment, reducing financial risks and providing more stable income. This capital allows farmers to plan and invest confidently, knowing that they have a committed base of customers before planting even begins. Consumers benefit by receiving fresh, locally grown produce that is often picked at peak ripeness, ensuring higher nutritional value and better taste compared to items that may have traveled long distances.

CSAs are also environmentally beneficial, as they minimize food miles and reduce the need for energy-intensive long-haul transportation and refrigerated storage. Local distribution reduces fuel consumption and emissions associated with conventional food supply chains, contributing to a smaller carbon footprint. Studies indicate that CSAs foster biodiversity on farms by encouraging crop rotation and the cultivation of heirloom and diverse crop varieties, which are often neglected in monoculture farming practices (Galt et al., 2019). This crop diversity not only enhances soil health but also promotes a more resilient ecosystem, as varied plant species can attract beneficial insects and reduce the need for chemical pesticides.

Another significant benefit of CSAs is the relationship they cultivate between farmers and consumers. By involving community members in the agricultural process, CSAs help foster a deeper understanding of and appreciation for the work that goes into food production. Some CSA programs even invite members to participate in farm events, volunteer days, or harvest festivals, which build a sense of community around food and agriculture. This direct connection between consumers and farmers nurtures transparency and trust, empowering consumers to make informed choices about their food while providing farmers with a supportive community of stakeholders who value their work and produce.

Urban Farming

Urban farming is an innovative solution to localizing food production within densely populated cities, where access to traditional farmland is limited. Urban farms use underutilized spaces—such as small plots, rooftops, abandoned lots, and even vertical walls—to grow vegetables, herbs, and sometimes fruits directly within city limits. This localized production model minimizes food miles and reduces the environmental costs associated with transportation, making it a sustainable option for urban areas. By growing food close to the consumer, urban farms can significantly decrease the need for fuel-intensive logistics, storage, and refrigeration, directly cutting down on carbon emissions.

Urban farming also promotes sustainable agricultural practices that align with environmentally responsible food production. Many urban farms incorporate composting to recycle organic waste, which enriches the soil and reduces the need for synthetic fertilizers. Rainwater harvesting systems are often used to supplement irrigation, conserving water resources and reducing dependence on municipal water supplies. Additionally, urban farms frequently employ organic pest control methods, such as introducing beneficial insects or using natural pest repellents, which further minimizes chemical runoff and protects urban ecosystems (Despommier, 2010). These practices contribute to the creation of green spaces in urban environments, improving air quality and enhancing biodiversity within cities.

In addition to environmental benefits, urban farming addresses social issues like food insecurity and unequal access to fresh produce. In many urban areas, particularly in underserved neighborhoods, residents lack access to affordable, fresh, and nutritious food. These areas, often referred to as "food deserts," are marked by a scarcity of grocery stores or farmers' markets and an abundance of fast-food outlets and convenience stores. By establishing urban farms and community gardens within these neighborhoods, cities can offer residents fresh produce at affordable prices, improving dietary options and health outcomes. Programs that distribute food from urban farms directly to community members or local food banks help to alleviate food insecurity while fostering community empowerment and resilience.

Though urban farming requires investment in infrastructure, such as irrigation systems, greenhouses, and vertical growing systems, it has proven to be a viable model for localized food production in cities across the United States, including New York, Chicago, and Detroit. In New York, for example, the organization Brooklyn Grange operates several rooftop farms, producing vegetables, herbs, and honey that are sold directly to local consumers and restaurants. Similarly, Chicago's Urban Growers Collective transforms vacant lots into productive agricultural spaces, supplying fresh produce to nearby neighborhoods and providing educational programs in food production, nutrition, and sustainable practices. In Detroit, urban farming initiatives such as Keep Growing Detroit have revitalized vacant lots, transforming them into productive farms that support food security, health, and economic development within the community.

Both community-supported agriculture and urban farming offer powerful models for creating resilient, sustainable, and localized food systems. By reducing food miles, promoting biodiversity, and building connections between producers and consumers, these initiatives help address the environmental, economic, and social challenges of the modern food system. CSAs provide consumers with an opportunity to support local farms, reduce their carbon footprint, and engage in a community-focused food system, while urban farming initiatives bring fresh food to underserved areas, empower local communities, and transform urban landscapes. Together, these localized food systems demonstrate the potential for innovative, community-centered approaches to reshape food production and consumption for a more sustainable future.

Challenges and Solutions for Localizing Food Systems

While the environmental and community benefits of local food systems are well-documented, there are several critical challenges that need to be addressed to make these systems viable, scalable, and resilient. By tackling these obstacles, local food networks can better meet consumer demand, support producers, and compete with large-scale, conventional food distribution systems.

Seasonality and Year-Round Production

One of the primary challenges facing local food systems is seasonality. Locally grown foods are often limited to what can be produced during specific growing seasons, which restricts the variety and availability of fresh produce during certain times of the year, especially in regions with harsh winters. Consumers accustomed to the continuous availability of imported produce may find it difficult to adapt to the limited selection that characterizes local and seasonal offerings. To address this issue, some local food systems are investing in greenhouse and indoor farming technologies that create controlled environments for year-round production. Greenhouses, when equipped with energy-efficient technologies like LED lighting and climate control, allow farmers to extend the growing season and produce crops beyond typical seasonal limitations. Indoor farming techniques such as hydroponics and vertical farming are especially promising, as they enable food production with minimal resource use in terms of water and land area, allowing for the cultivation of leafy greens, herbs, and certain fruits even in urban settings (Al-Kodmany, 2018). Although these technologies hold great promise, they require substantial initial investments in infrastructure, energy, and skilled labor, which can present financial challenges for small-scale farmers.

Infrastructure and Distribution Needs

Another significant challenge for local food systems is the need for infrastructure to efficiently process, store, and distribute fresh produce. Small-scale farmers often lack access to essential facilities such as refrigerated storage, commercial kitchens, and distribution networks, which limits their ability to handle high volumes, meet food safety standards, and reach a broader market. Without adequate storage and processing facilities, local producers may face losses due to spoilage, and their reach remains limited to immediate, local markets.

Regional food hubs are emerging as a solution to this infrastructure gap by offering centralized facilities that aggregate, process, and distribute products from multiple local farms. These hubs provide shared resources, such as refrigerated storage and packaging equipment, which increase efficiency and reduce costs for small producers. Food hubs also facilitate connections between local farmers and institutional buyers, such as restaurants, schools, and grocery stores, enabling farmers to expand their customer base and achieve economies of scale (Matson et al., 2013). By bridging the infrastructure gap, food hubs play a crucial role in supporting the growth and scalability of local food systems, making them more competitive with conventional, global food networks.

Land and Labor Costs

Land and labor costs are formidable barriers to expanding local food systems, particularly in urban and peri-urban areas where real estate values are high. As cities grow, agricultural land near urban centers often becomes repurposed for housing or commercial development, making it increasingly difficult to sustain or expand urban farming initiatives. Farmland preservation initiatives and zoning policies that protect agricultural land near cities are essential for maintaining local food production in these high-demand areas. Policies that designate

107

agricultural zones or incentivize the conservation of farmland can prevent urban sprawl from encroaching on valuable agricultural spaces, thereby securing local food sources for future generations.

Labor shortages and the high costs of skilled farm labor are additional challenges. Many small-scale and urban farms struggle to attract and retain workers due to the physically demanding nature of farm work and the relatively low wages it often commands. Programs that support fair wages, training, and benefits for farmworkers can help address these labor challenges and make local agriculture more sustainable in the long term. Some governments provide subsidies or grants to support local food systems, which can offset costs for land acquisition, labor, and infrastructure development. These subsidies encourage farmers to continue working the land and help keep local food production viable in regions with high operational costs.

Consumer Education and Awareness

Consumer education and awareness are critical for the success of local food systems, as purchasing behaviors directly influence the demand for locally sourced food. Many consumers are unaware of the environmental and economic benefits of purchasing local products, and convenience often drives purchasing decisions. Without clear information, consumers may not recognize the broader impact of supporting local farmers, and they may overlook local options in favor of more readily available or convenient imported goods.

Public awareness campaigns and labeling systems that highlight locally grown products are effective tools for encouraging consumers to support local food. For instance, labels indicating "locally sourced" or "farm-to-table" products can help consumers identify and choose foods that align with their values. Additionally, some states and municipalities are implementing "buy local" programs that incentivize purchasing from local farmers and increase the visibility of regional foods in grocery stores and restaurants. These programs may offer rewards, discounts, or loyalty points for customers who prioritize local purchases, helping to foster a market for local produce (Coleman-Jensen et al., 2019). By educating consumers about the environmental, economic, and community benefits of purchasing local food, these initiatives help build support for local food systems and foster a more resilient and sustainable food network.

Building a Resilient Local Food System

To successfully build resilient and competitive local food systems, a combination of technological, financial, policy-based, and educational strategies is essential. Addressing seasonality through indoor farming and greenhouses, developing regional food hubs to bridge infrastructure gaps, implementing land-use policies to protect farmland, and educating consumers on the benefits of local food are all part of creating a robust local food economy. Together, these solutions allow local food systems to better serve both producers and consumers, reduce environmental impacts, and contribute to stronger, healthier communities.

By tackling these challenges head-on, local food systems can grow to become an integral part of a sustainable food future, offering an alternative to globalized supply chains and supporting environmental and economic resilience at the community level.

Chapter 13

Technological Advances in Reducing Food Chain Emissions

Technological advancements are increasingly leveraged to reduce greenhouse gas (GHG) emissions throughout the food supply chain, offering solutions that address emissions from production, processing, transportation, and waste management. These innovations range from optimizing cold storage and logistics to harnessing the latest in digital technology, each contributing to a more efficient, less carbon-intensive food system.

Energy-Efficient Refrigeration and Logistics Systems

One of the most impactful advancements in reducing emissions within the food chain is the adoption of energy-efficient refrigeration and logistics systems. Refrigeration accounts for a significant portion of energy use in food storage and transportation, particularly for perishable goods. To address this, improved refrigeration technologies, such as smart refrigeration systems equipped with sensors and automated controls, have been developed to maintain optimal temperatures while using less energy. These systems monitor food storage conditions in real-time, adjusting temperatures as needed and alerting operators to potential issues such as mechanical failures or temperature fluctuations.

This precise temperature control not only lowers energy consumption but also extends the shelf life of products, helping to reduce food spoilage and waste, which in turn reduces the emissions associated with producing and disposing of food that might otherwise be lost (Jones et al., 2017).

Additionally, some systems are exploring renewable energy sources for powering cold storage facilities. Solar-powered refrigeration units, for example, are being tested in areas with high sunlight exposure, allowing facilities to maintain low temperatures with minimal carbon emissions. Other advances include phase-change materials and vacuum-insulated panels, which improve insulation and reduce the energy demand for refrigeration. Combined, these technologies contribute to significant carbon savings throughout the food cold chain.

Blockchain Technology in Food Logistics

Blockchain technology has emerged as a transformative tool in the food industry by enhancing transparency and traceability within the supply chain. Blockchain securely records each stage of a product's journey, from farm to consumer, by using an immutable digital ledger. By providing precise tracking information on each batch of food, blockchain allows supply chain operators to efficiently manage inventory, plan transportation schedules, and make informed decisions on storage needs. This level of tracking and transparency reduces the need for over-storage and minimizes delays in distribution, thereby lowering the carbon footprint associated with food waste and inefficient transportation practices. For example, blockchain enables companies to pinpoint exactly where a product is in transit or storage, allowing for better planning to avoid unnecessary refrigeration or waste (Kamilaris et al., 2019).

The use of blockchain also has significant implications for food safety. In cases of contamination, blockchain allows for rapid tracing of affected batches, enabling faster recalls and reducing the volume of food discarded as a precautionary measure. This precision decreases the amount of food lost to waste and the associated GHG emissions from growing, processing, and disposing of those products. Major food companies, including Walmart and IBM, have already integrated blockchain into their supply chains to enhance traceability and sustainability, setting a precedent for the broader industry.

Electric and Hybrid Delivery Trucks

Transportation is a major contributor to emissions in the food supply chain, especially for long-haul and last-mile deliveries. The adoption of electric and hybrid delivery trucks has the potential to significantly reduce the environmental impact of food transport. Electric vehicles (EVs) and hybrid trucks produce far fewer emissions than traditional diesel-powered trucks, particularly in urban areas where traffic congestion can lead to prolonged idling and increased emissions. These vehicles also contribute to reducing air pollution, a major concern in cities with high population densities (Johnson & Martin, 2018).

Many logistics companies and food retailers are now incorporating electric trucks into their fleets, especially for last-mile deliveries. This shift to EVs is driven by both regulatory pressures and consumer demand for more sustainable transportation options. For instance, some urban areas have introduced low-emission zones that restrict access to fossil fuel-powered vehicles, incentivizing businesses to adopt cleaner technologies. Additionally, advances in battery technology are extending the range of electric vehicles, making them increasingly viable for longer routes and reducing range anxiety. Companies such as Amazon and UPS are investing in electric delivery fleets, recognizing the long-term economic and environmental benefits of reduced fuel costs and lower emissions.

Artificial Intelligence and Machine Learning in Logistics Optimization

Artificial intelligence (AI) and machine learning are playing a growing role in optimizing logistics within the food supply chain. By analyzing vast amounts of data on routes, traffic patterns, and fuel consumption, AI algorithms can help companies design the most efficient delivery routes, reducing travel time and fuel use. Machine learning models are also used to forecast demand and streamline inventory management, ensuring that only the necessary amount of food is produced, transported, and stored at any given time. This reduces waste and minimizes emissions associated with overproduction and surplus inventory (Smith et al., 2020).

For example, AI-driven demand forecasting can anticipate peak periods for certain food items, allowing suppliers to adjust production and transportation schedules to meet demand without creating surplus. By avoiding the pitfalls of overproduction and last-minute rush logistics, companies can prevent unnecessary emissions while meeting consumer needs more sustainably. Additionally, AI in warehouse management systems optimizes product placement and retrieval, reducing the energy consumption associated with food handling and storage.

Drones and Autonomous Vehicles in Agriculture and Distribution

Drones and autonomous vehicles are revolutionizing food production and distribution, particularly in agriculture and logistics. In agriculture, drones equipped with sensors and cameras allow for precision monitoring of crop health, soil conditions, and water needs. By providing farmers with real-time data, drones enable more precise application of water, fertilizers, and pesticides, significantly reducing resource use and emissions associated with excessive inputs (Zhang et al., 2018).

In distribution, autonomous delivery vehicles, including drones, are being tested for last-mile deliveries in urban areas. Companies like Amazon and UPS have trialed drones to deliver packages quickly and efficiently, particularly in locations that are hard to reach by conventional trucks. Autonomous vehicles are also being piloted for short-haul transport in warehouse complexes and controlled environments. These technologies have the potential to reduce emissions by cutting down on fuel consumption and improving the speed and efficiency of the delivery process.

Together, these technological advancements are creating a more efficient and sustainable food supply chain. By leveraging smart refrigeration, blockchain for traceability, electric vehicles, AI in logistics, and autonomous delivery, the food industry is transforming to meet the dual challenges of reducing emissions and addressing global demand. As these technologies continue to develop and become more widely accessible, they are likely to play a critical role in building a sustainable, low-carbon future for food production and distribution.

The Rise of Lab-Grown Meat, Vertical Farming, and Precision Agriculture

Emerging food production technologies are revolutionizing sustainable agriculture by enhancing resource efficiency, lowering emissions, and reshaping how food is produced. These innovations are tackling some of the most significant environmental challenges associated with traditional farming practices, including high water and land usage, methane emissions, and pesticide runoff. Among these technologies, lab-grown meat, vertical farming, and precision agriculture stand out as promising solutions that could help reshape food production for a sustainable future.

Lab-Grown Meat

Lab-grown meat, also known as cultured or cell-based meat, has generated considerable interest as an alternative to traditional animal agriculture. Unlike conventional meat, which relies on raising and slaughtering animals, lab-grown meat is produced by culturing animal cells in a controlled environment. Cells are harvested from animals and cultivated with nutrients and growth factors in bioreactors, where they replicate to form muscle tissue, the main component of meat. This method requires significantly fewer natural resources, such as land and water, and emits far less greenhouse gases (GHGs) compared to traditional meat production, which is a significant contributor to methane emissions from livestock (Chriki & Hocquette, 2020).

The environmental benefits of lab-grown meat are substantial. Research indicates that lab-grown meat could reduce land use by over 90% and water use by up to 50% compared to conventional beef production. Livestock farming occupies vast areas of land and drives deforestation, while lab-grown meat requires minimal space. With a smaller ecological footprint, lab-grown meat addresses many of the environmental concerns linked to livestock farming, including habitat destruction and methane emissions. Moreover, lab-grown meat has the potential to reduce the ethical and welfare issues associated with animal farming, as it minimizes the need to raise and slaughter animals. Although lab-grown meat is still in the early stages of commercialization, companies like Memphis Meats and Mosa Meat are making progress, with production costs gradually declining as the technology advances.

However, challenges remain. Scaling up lab-grown meat production is costly, and creating a product that mimics the texture and flavor of traditional meat requires continuous refinement. Public acceptance also poses a hurdle, as consumers may be hesitant to adopt lab-grown meat due to its novel nature. Nevertheless, as production methods improve and prices decrease, lab-grown meat could play a crucial role in reducing the environmental impact of protein production while meeting the global demand for meat.

Vertical Farming

Vertical farming is an innovative solution for sustainable food production, particularly suited for urban areas with limited agricultural space. In vertical farms, crops are grown in stacked layers within controlled-environment agriculture (CEA) systems, which regulate temperature, humidity, and light to optimize plant growth. By utilizing LED lighting and hydroponic or aeroponic growing methods, vertical farms use up to 95% less water than traditional soil-based agriculture and do not require pesticides. These controlled systems make vertical farming highly resource-efficient and sustainable, enabling year-round food production with minimal environmental impact (Al-Kodmany, 2018).

Vertical farms are especially advantageous in urban settings, where they can reduce "food miles"—the distance food travels from production to consumer—by situating food production closer to where people live. Companies such as AeroFarms and Plenty are at the forefront of vertical farming innovation, developing large-scale facilities that supply fresh produce to urban populations. These companies have demonstrated that vertical farms can yield significantly higher outputs per square meter than conventional farms, making them an efficient solution for meeting urban food demand. By producing food locally, vertical farms reduce the need for transportation and refrigeration, further lowering emissions. Additionally, since vertical farms operate in closed-loop systems, they can recycle water and nutrients, minimizing waste and runoff pollution.

Despite its benefits, vertical farming faces economic challenges. Establishing and maintaining vertical farms requires substantial upfront investment in technology, infrastructure, and energy, especially for LED lighting. Energy consumption for lighting and climate control is a significant operational cost, and efforts are underway to develop energy-efficient systems to make vertical farming more sustainable and economically viable. Nevertheless, as renewable energy sources become more accessible and energy-efficient technologies improve, vertical farming has the potential to contribute meaningfully to sustainable food production in urban environments.

Precision Agriculture

Precision agriculture is a data-driven approach that employs advanced technologies, such as satellite imagery, sensors, and GPS technology, to optimize crop management. By using real-time data to monitor soil conditions, moisture levels, and crop health, precision agriculture enables farmers to make informed decisions about resource allocation. This level of precision allows for targeted application of water, fertilizers, and pesticides, reducing waste and minimizing environmental impact. Studies show that precision agriculture can decrease water usage by up to 20% and reduce pesticide use by 30%, significantly lowering the carbon footprint of farming operations (Zhang et al., 2018).

In addition to environmental benefits, precision agriculture improves crop yields and resource efficiency, which are essential for food security. Automated tractors and drones can survey large fields, identify areas needing intervention, and apply inputs with high accuracy. For example, drones equipped with infrared cameras can detect stressed plants, enabling farmers to address specific issues without treating entire fields.

Soil sensors provide continuous data on nutrient levels, allowing farmers to apply fertilizers only where needed, thereby preventing nutrient runoff into nearby waterways.

While precision agriculture offers numerous benefits, its adoption is limited by the high cost of technology and the need for technical expertise. Small-scale farmers, particularly in developing regions, may struggle to afford precision equipment and software, which can limit the widespread application of these innovations. However, as the costs of sensors, drones, and data analytics decrease, precision agriculture is becoming more accessible, offering farmers the opportunity to improve productivity while minimizing environmental impact.

Lab-grown meat, vertical farming, and precision agriculture represent transformative solutions in the quest for sustainable food production. By reducing resource use, emissions, and waste, these technologies address some of the most pressing environmental issues associated with agriculture. Although challenges remain, continued innovation and investment can help overcome these obstacles, paving the way for a more resilient and sustainable food system. As these technologies evolve, they hold the potential to reshape global agriculture, offering new ways to meet food demand while preserving natural resources and reducing the ecological footprint of food production.

Economic and Environmental Impacts of Sustainable Innovations

The economic and environmental impacts of sustainable innovations in the food industry are transformative, offering extensive benefits that span multiple sectors and address some of the most pressing global challenges. Technologies designed to reduce emissions, enhance resource efficiency, and minimize waste hold the potential not only to improve environmental sustainability but also to drive economic growth by lowering production costs, increasing yields, and stabilizing food prices over time. This dual impact of sustainability and profitability is shaping a new landscape for the food industry, fostering resilience and supporting food security worldwide.

Economic Benefits and New Opportunities

Sustainable innovations like precision agriculture, vertical farming, and lab-grown meat introduce new business models and investment opportunities in the growing food tech sector, spurring economic growth and job creation in both urban and rural areas. For instance, precision agriculture reduces costs for farmers by optimizing the application of inputs such as water, fertilizers, and pesticides. By using sensors, drones, and GPS technology to target specific areas, precision agriculture enables farmers to apply resources where they are most needed, enhancing crop yields while lowering input costs. This targeted approach not only reduces waste but also increases profitability for farmers, improving their long-term economic stability and contributing to local food security (Specht et al., 2019).

Similarly, vertical farming and lab-grown meat represent promising new markets that attract investors and drive economic growth. Vertical farming, for example, allows food to be grown in densely populated urban centers, reducing transportation costs and enabling fresher produce to reach consumers more quickly. This model creates jobs in agriculture, technology, and logistics within urban settings, making food production more integrated with the local economy. Lab-grown meat also presents an emerging market that is attracting significant investment. As production techniques improve and costs decrease, lab-grown meat could potentially become a mainstream source of protein, offering economic opportunities across the food tech sector, from production and distribution to consumer sales and marketing. By creating sustainable, scalable food production methods, these innovations support economic resilience and provide new avenues for economic development in the food industry.

Environmental Benefits and Climate Impact

The environmental impacts of sustainable innovations are equally significant. By minimizing resource use, reducing emissions, and preventing waste, these technologies contribute to mitigating climate change, conserving natural resources, and decreasing pollution. Lab-grown meat, for instance, provides a way to produce protein without the environmental downsides of traditional livestock farming, which is associated with deforestation, methane emissions, and high water consumption. Research suggests that lab-grown meat could reduce methane emissions by up to 90% compared to conventional beef, providing a viable alternative for environmentally conscious consumers (Chriki & Hocquette, 2020).

Vertical farming and precision agriculture further contribute to reducing the ecological footprint of food production by conserving land, water, and energy. Vertical farming operates in controlled environments that use up to 95% less water than traditional soil-based agriculture, as water is recirculated within hydroponic or aeroponic systems. Additionally, vertical farming eliminates the need for pesticides and herbicides, reducing chemical runoff that can pollute waterways. Precision agriculture, meanwhile, reduces soil degradation and minimizes fertilizer and pesticide use, resulting in less nitrogen runoff and fewer harmful environmental effects. Together, these innovations support global efforts to preserve natural resources, limit pollution, and adapt to the environmental constraints of a growing global population.

Challenges and the Role of Policy Interventions

Despite the immense potential of these technologies, several challenges remain, particularly regarding cost and consumer acceptance. The initial investment required to implement advanced technologies like vertical farming and precision agriculture can be prohibitive for small-scale farmers and startup businesses, who may lack the capital necessary to invest in high-tech equipment and infrastructure. Additionally, ongoing operational costs, such as energy use for vertical farms and maintenance for precision agriculture systems, can add to the financial burden, making these technologies difficult to scale without financial support.

Policy interventions are essential to bridging this gap. Subsidies, grants, and research funding can make sustainable technologies more accessible to a wider range of producers, enabling smaller farms and emerging businesses to adopt these practices. For example, government grants for research and development can drive innovation in renewable energy sources for vertical farms, making them more affordable and sustainable. Tax incentives and subsidies for precision agriculture equipment can reduce initial costs for farmers, helping them transition to more efficient and environmentally friendly methods. Additionally, public funding for training and technical support can equip farmers with the skills needed to use these new technologies effectively.

Consumer acceptance of lab-grown meat and other novel foods also presents a significant hurdle. Cultural and psychological factors influence how consumers perceive lab-grown meat, with some expressing hesitation or ethical concerns about consuming products that are created in a laboratory rather than naturally raised. To support the wider adoption of these innovations, public awareness campaigns and educational initiatives can help shift perceptions, emphasizing the environmental and ethical benefits of lab-grown meat and other sustainable alternatives. Regulatory frameworks that ensure the safety, transparency, and quality of lab-grown and other innovative food products can also foster trust, helping consumers feel more confident in choosing sustainable options.

In summary, sustainable innovations in the food industry—including lab-grown meat, vertical farming, precision agriculture, and blockchain—are essential for reducing emissions, conserving resources, and building resilient food systems. These technologies offer transformative economic and environmental benefits that pave the way for a sustainable future, addressing critical challenges such as climate change, resource scarcity, and global food security. As these innovations continue to evolve and scale, they hold the potential to reshape the food industry, enabling a more sustainable, resilient, and resource-efficient global food system that can meet the needs of a growing population while protecting the planet.

Chapter 14

Policy and Global Perspectives

Global food systems are at the nexus of environmental sustainability and economic stability, with their carbon footprints shaped by an intricate web of policies, trade agreements, and collaborative initiatives. From local regulations promoting organic farming to international frameworks addressing greenhouse gas (GHG) emissions, policies play a pivotal role in directing the environmental and economic trajectory of food production and distribution. By examining the policies that drive sustainable agricultural practices, reduce food waste, and foster global cooperation, this chapter delves into the mechanisms that underpin efforts to create a low-carbon, resilient food system. Through case studies and analysis, we explore how targeted regulations, incentives, and international collaboration can transform food systems into engines of sustainability and equity.

Policies for Sustainable Agriculture and Emissions Reduction

Government policies are essential for incentivizing the adoption of sustainable farming practices that reduce the environmental impact of food production. One prominent example is the Common Agricultural Policy (CAP) of the European Union (EU), which provides financial support to farmers who implement environmentally friendly agricultural practices. Under CAP, farmers can receive subsidies for actions such as crop rotation, reduced pesticide and fertilizer use, and practices that promote soil health and biodiversity. For instance, the "greening" measures introduced in CAP reforms aim to make farming more environmentally sustainable by encouraging practices that reduce GHG emissions and support biodiversity, such as planting cover crops and maintaining permanent pastures (European Commission, 2018). These measures contribute to lowering the carbon footprint of farming while maintaining ecosystem services that are essential for sustainable food production. Additionally, the EU has strengthened its focus on promoting organic farming through direct payments to farmers who switch to organic practices, further aligning agricultural policies with environmental goals.

In the United States, similar efforts are underway through programs like the Environmental Quality Incentives Program (EQIP), which offers both financial and technical assistance to farmers and ranchers adopting practices that reduce their environmental impact. This program provides incentives for practices such as precision agriculture, conservation tillage, and the adoption of renewable energy solutions like solar panels on farms. EQIP is designed to support efforts to reduce emissions, conserve water, and protect soil health, helping farmers transition to more sustainable operations while maintaining productivity (USDA, 2020). By encouraging these practices, EQIP plays a crucial role in lowering the carbon footprint of U.S. agriculture while promoting resource conservation.

In addition to general environmental incentives, several U.S. states have introduced specific regulations targeting methane emissions from the agricultural sector. California, for instance, has introduced comprehensive methane reduction regulations that focus on reducing emissions from livestock operations, particularly dairy farms, which are major sources of methane—a potent greenhouse gas. These regulations incentivize farmers to adopt technologies such as anaerobic digesters, which capture methane produced by manure and convert it into renewable energy. This not only helps mitigate methane emissions but also provides a sustainable source of power for farms, reducing reliance on fossil fuels and contributing to California's broader climate goals

(California Department of Food and Agriculture, 2020). Such policies demonstrate how targeted regulations can address specific sources of emissions within agriculture while creating economic incentives for farmers to invest in cleaner technologies.

Food Waste Policies and Their Environmental Impact

Government policies that target food waste play a significant role in reducing emissions associated with discarded food. Globally, food waste contributes substantially to GHG emissions, particularly methane, which is produced when food decomposes in landfills without oxygen. To combat this, governments are implementing policies that encourage the reduction of food waste at every stage of the food chain, from production to consumption. France stands out as a pioneer in this area. In 2016, France became the first country to pass legislation requiring supermarkets to donate unsold but edible food to charities instead of discarding it. This law has led to a dramatic decrease in food waste in the retail sector, providing a social benefit by redirecting surplus food to those in need while also reducing the methane emissions that would otherwise result from landfilled food (Mourad, 2016).

In addition to encouraging food donation, the French law mandates that supermarkets with over 400 square meters of retail space must also prevent food waste by ensuring unsold products are either repurposed for animal feed or composted. The policy has not only reduced the environmental impact of food waste but has also set a strong example for other countries looking to curb food waste and its associated emissions. Several nations, including the UK, Canada, and the United States, have followed suit by introducing similar measures, demonstrating the growing importance of food waste reduction as a key strategy for mitigating climate change.

Policies like these highlight the power of targeted regulation to address both social and environmental challenges within the food system. Reducing food waste not only addresses hunger and food insecurity but also contributes to mitigating climate change by reducing methane emissions from landfills. In addition to regulatory approaches, public awareness campaigns have been launched in many countries to educate consumers and businesses about the importance of reducing food waste and implementing sustainable food management practices. These efforts contribute to a broader cultural shift toward more responsible consumption and disposal of food.

Integrating Policy Approaches for Greater Impact

The interplay between agricultural practices, food waste, and emissions reduction underscores the importance of integrated policy approaches that link different stages of the food chain. Policies that encourage sustainable farming, waste reduction, and the adoption of low-carbon technologies can create synergies that maximize environmental benefits. For example, integrating food waste reduction policies with sustainable farming incentives could help to create a circular economy in the food system, where food waste is minimized and organic materials are recycled back into the soil, enhancing agricultural productivity and reducing reliance on synthetic inputs.

Furthermore, international collaboration and alignment of food-related policies can drive progress on a global scale. As food production and consumption are inherently global, aligning policies across borders can help ensure that sustainability goals are met and that countries share knowledge and resources to reduce food-related emissions. Through international frameworks such as the Paris Agreement and the UN Sustainable Development Goals (SDGs), nations can work together to create a unified global approach to tackling food system emissions, reducing waste, and achieving a sustainable, low-carbon food future.

Government policies play a crucial role in shaping the environmental footprint of the food supply chain. By providing incentives for sustainable farming practices, addressing methane emissions, and promoting food waste reduction, policymakers can drive substantial reductions in greenhouse gas emissions and help create more resilient food systems. Case studies from countries such as the EU, the U.S., and France illustrate how targeted regulations and innovative policies can simultaneously address environmental concerns and promote economic growth. As nations continue to align their agricultural and food policies with sustainability goals, the global food system will become increasingly capable of meeting the challenges of climate change, food security, and resource conservation.

The Role of International Cooperation in Reducing Food-Related Emissions

Given the global nature of food production and distribution, international cooperation is essential for effectively addressing the carbon footprint of the food chain and ensuring sustainable food systems. The complexities of the global food market mean that emissions reductions and sustainability improvements require collaborative efforts across countries and organizations. To facilitate these efforts, global organizations such as the United Nations (UN), the Food and Agriculture Organization (FAO), and the Intergovernmental Panel on Climate Change (IPCC) play critical roles in establishing international guidelines, setting ambitious targets, and sharing best practices for reducing food-related emissions. These organizations provide a framework for countries to align their actions with global sustainability goals, fostering a unified approach to addressing environmental challenges in the food sector.

One of the most comprehensive frameworks guiding these efforts is the UN's Sustainable Development Goals (SDGs). The SDGs emphasize the need for responsible consumption and production practices through Goal 12, which promotes actions to reduce food waste, improve agricultural efficiency, and promote sustainable resource use. This goal provides a clear target for nations to work toward sustainable food systems that can meet the needs of a growing population without compromising environmental health (UN, 2015). The SDGs encourage countries to create policies that minimize the environmental impact of food production and prioritize equitable resource distribution, making sustainable development a core part of international cooperation in the food sector.

The Paris Agreement and Agricultural Emissions

The Paris Agreement represents a landmark achievement in international cooperation on climate change, and its implications for food system sustainability are substantial. Under this agreement, countries commit to national climate action plans, known as Nationally Determined Contributions (NDCs), which outline specific targets for reducing emissions and adapting to climate impacts. Recognizing that agriculture is a significant contributor to global greenhouse gas (GHG) emissions—accounting for roughly a quarter of total emissions, including land use changes—many countries have integrated agricultural emissions reductions into their NDCs. These commitments often include measures to increase agricultural resilience, reduce methane and nitrous oxide emissions, and implement sustainable farming practices (IPCC, 2019).

Through the Paris Agreement framework, countries collaborate to share knowledge, technological advancements, and resources to accelerate progress toward emissions reductions. For example, countries with advanced capabilities in precision agriculture, soil carbon sequestration, and methane reduction technologies are encouraged to share these innovations with other nations. This exchange of expertise fosters a collaborative approach to mitigating climate impacts and enhances the capacity of all countries to address the carbon footprint of their food systems. By working together to meet the targets set by the Paris Agreement, participating nations not only reduce their own emissions but also contribute to a collective effort to stabilize global temperatures and build resilient food systems.

The Role of International Trade Agreements in Sustainability

International trade agreements are another key mechanism through which countries can work together to reduce the food chain's carbon output. Trade policies that support the exchange of sustainably produced goods create economic incentives for low-carbon products and encourage producers worldwide to adopt sustainable practices. One significant example of this approach is the EU-Mercosur Trade Agreement, which includes provisions for environmental standards in agriculture. This agreement, which facilitates trade between the European Union and Mercosur (a South American trade bloc that includes Argentina, Brazil, Paraguay, and Uruguay), includes specific requirements for reducing deforestation and promoting sustainable farming practices. In exchange for greater market access to the EU, South American countries are encouraged to adopt environmental protections and low-emission practices in their agricultural sectors (European Commission, 2019).

By setting sustainability standards as part of trade agreements, countries can promote a global standard for food production that values environmental stewardship and resource conservation. Such agreements incentivize countries to pursue sustainability goals by rewarding sustainable practices with access to lucrative markets. Trade policies that emphasize sustainable practices also encourage companies and producers to reduce emissions throughout their supply chains, fostering a global shift toward lower-carbon production processes. In this way, international trade agreements not only promote economic growth but also support a coordinated global effort to address the environmental impact of food production.

Collaborative Initiatives and Knowledge Sharing

In addition to formal trade agreements, countries collaborate on sustainability initiatives through international partnerships and programs. One example is the 4 per 1000 Initiative, launched at the COP21 climate conference in Paris. This initiative encourages countries to increase soil carbon storage by 0.4% annually, which can improve soil health, boost agricultural productivity, and reduce atmospheric CO_2. By promoting practices like no-till farming, cover cropping, and agroforestry, the 4 per 1000 Initiative provides a platform for countries to share soil carbon sequestration strategies, offering a practical approach to reduce emissions in agriculture and mitigate climate change (Minasny et al., 2017).

The Global Alliance for Climate-Smart Agriculture (GACSA) is another collaborative platform that brings together governments, businesses, and civil society organizations to promote climate-smart agriculture practices. GACSA's focus is on increasing agricultural resilience, improving resource efficiency, and reducing emissions in farming. By creating a global network of climate-smart agriculture advocates,

GACSA facilitates the sharing of best practices and supports countries in implementing policies that promote sustainable food systems. This collaborative approach helps countries, particularly those in developing regions, access the resources and knowledge necessary to enhance food security while minimizing environmental impact (FAO, 2018).

Addressing Challenges and Moving Forward

While international cooperation has made significant progress in promoting sustainable food systems, challenges remain. The economic and political interests of different countries can create conflicts, as nations with high agricultural emissions may face pressure to change practices that are economically advantageous. Additionally, the implementation of sustainability commitments varies widely across countries due to differences in financial resources, technological capacity, and regulatory frameworks. To address these challenges, ongoing dialogue and negotiation are essential, as is the provision of financial support to lower-income countries. Programs like the Green Climate Fund provide critical funding for climate adaptation and mitigation projects, helping countries with limited resources to implement sustainable food practices.

Moving forward, continued international cooperation will be crucial to addressing the complex environmental and economic challenges facing the global food system. The success of frameworks like the Paris Agreement, the SDGs, and collaborative initiatives hinges on nations' commitment to prioritizing sustainability, investing in technological innovations, and sharing resources equitably. Through collective action, countries can build a resilient, low-carbon food system that meets the needs of a growing global population while protecting the planet.

Case Studies of Countries with Successful Food Sustainability Models

Certain countries have implemented highly effective food sustainability models that showcase best practices for reducing emissions, promoting environmental responsibility, and achieving long-term resilience in the food system. Through targeted policies, technological innovations, and widespread public engagement, these countries offer examples that other nations can look to for guidance in building sustainable food systems.

Denmark: A Leader in Organic Farming

Denmark stands out for its commitment to organic farming, with over 10% of its agricultural land dedicated to organic production—a proportion that is among the highest in the world. This emphasis on organic farming is largely driven by Denmark's Organic Action Plan, which aims to increase organic production through a combination of subsidies, farmer education, and market support. Under this plan, farmers who transition to organic practices receive financial incentives to offset initial costs, along with access to resources and training to meet organic certification standards. The plan also invests in the development of local markets for organic products, creating an economically viable ecosystem for organic farming (Organic Denmark, 2019).

The environmental benefits of Denmark's model are substantial. Organic farming avoids synthetic fertilizers and pesticides, which are not only significant contributors to greenhouse gas (GHG) emissions but also detrimental to soil health and biodiversity. By eliminating synthetic inputs, organic farming in Denmark promotes soil health, reduces nitrogen runoff, and enhances biodiversity in agricultural landscapes. Additionally, organic methods contribute to carbon sequestration by enhancing soil organic matter. Denmark's Organic Action Plan has not only reduced emissions but has also inspired other EU countries to adopt similar goals, positioning organic farming as a key component of the EU's agricultural sustainability strategy.

Japan: Innovative Food Waste Management

Japan has developed an innovative model for managing food waste, with policies that address waste reduction at multiple stages of the food chain. Central to Japan's success is the Food Recycling Law, which mandates that businesses, including food manufacturers, retailers, and restaurants, recycle food waste rather than dispose of it. This law has spurred the creation of "eco-feed," where food waste is processed into animal feed, effectively repurposing waste as a resource for agriculture. Additionally, the Japanese government has launched public awareness campaigns and educational programs to inform citizens about proper food storage, consumption, and waste reduction practices, creating a cultural shift toward sustainable consumption (Japan Ministry of Agriculture, Forestry and Fisheries, 2020).

The Food Recycling Law has led to a 50% reduction in food waste since its implementation, demonstrating a successful model of waste reduction that aligns with circular economy principles. By repurposing food waste for animal feed, Japan reduces methane emissions that would have been generated if the food had decomposed in landfills. The eco-feed program also reduces reliance on imported feed, contributing to food security while minimizing the carbon footprint of livestock production. Japan's approach to food waste demonstrates how regulatory frameworks combined with public engagement can foster a sustainable food system and reduce environmental impact.

The Netherlands: Precision Agriculture and High-Tech Greenhouses

The Netherlands is a global leader in sustainable agriculture and emissions reduction, exemplified by its adoption of precision agriculture and high-tech greenhouses. Dutch farmers maximize yields through data-driven techniques that allow for precise management of inputs such as water, fertilizers, and pesticides. By employing advanced sensor technology, GPS systems, and real-time monitoring, precision agriculture in the Netherlands ensures that resources are used efficiently and only as needed, reducing waste and emissions. This approach minimizes runoff and nutrient leaching, thereby protecting surrounding ecosystems while optimizing crop health and productivity (van der Zee, 2017).

In addition to precision agriculture, the Netherlands has invested heavily in high-tech greenhouses that use closed-loop systems to recycle water and nutrients. These greenhouses operate with advanced climate control and LED lighting, enabling crops to grow with minimal energy and water use. By producing food year-round, these greenhouses also reduce the need for imports, minimizing transportation emissions. The high productivity and sustainability of the Dutch model have made the Netherlands the world's second-largest exporter of agricultural products, despite its small land area. This achievement illustrates how sustainable innovations can drive economic growth while preserving natural resources and reducing environmental impact.

Lessons from Global Models

The successful models implemented by Denmark, Japan, and the Netherlands demonstrate that targeted policies, innovative technologies, and cultural shifts can make significant strides toward sustainable food systems. Denmark's organic farming sector shows the potential of government support for sustainable agriculture practices that benefit both the environment and the economy. Japan's food waste policies highlight the importance of comprehensive waste management systems that engage both businesses and consumers in minimizing environmental impact. The Netherlands exemplifies how technological innovation in agriculture can increase efficiency and yield while reducing emissions and conserving resources.

By learning from these examples, other countries can adopt similar strategies to achieve sustainability in their own food systems. International cooperation and knowledge sharing are essential to replicating these successes globally. For instance, the principles of Japan's food recycling and eco-feed program could be adapted by other nations seeking to reduce food waste, while Denmark's organic action plan offers a model for promoting sustainable farming practices. These case studies emphasize that achieving a low-carbon food system is possible with the right combination of policies, innovation, and public commitment.

Chapter 15

Practical Steps for Consumers, Producers, and Policymakers

Consumers play a critical role in shaping the food system through their purchasing habits, dietary choices, and food waste management. As demand grows for sustainable food options, consumer behavior influences production practices, drives market trends, and even affects policy decisions. By adopting sustainable practices, individuals can significantly reduce their food carbon footprint and contribute to a more eco-friendly food chain, helping to mitigate climate change and conserve resources. With simple but impactful choices, consumers can encourage the food industry to prioritize sustainability and foster a more responsible food culture.

Opting for Plant-Based Foods

One of the most effective actions consumers can take to reduce their environmental impact is to incorporate more plant-based foods into their diets. Plant-based diets generally have a much lower carbon footprint than diets that are high in animal products, which require extensive resources and produce substantial greenhouse gas (GHG) emissions. Livestock farming is a major source of methane, a potent GHG, and studies have shown that beef production alone is responsible for around 60% of emissions from livestock, despite contributing less than 5% of total calories in the global diet (Poore & Nemecek, 2018). Reducing meat consumption, especially beef, has been shown to lower individual carbon footprints significantly.

Choosing plant-based foods also conserves water and land, as plant crops generally require less water and fewer land resources to produce the same amount of calories compared to animal products. For example, legumes and grains use a fraction of the water needed for beef production, making them more sustainable choices for daily consumption. By shifting to plant-based proteins, such as lentils, chickpeas, and tofu, consumers can make a substantial impact on water conservation and reduce the environmental strain caused by meat production.

Buying Seasonal and Local Foods

Another practical step for consumers is to buy seasonal and local foods. Locally sourced foods have shorter transportation distances, known as "food miles," which significantly reduces the emissions associated with transportation and refrigeration. When food is transported long distances, it often requires refrigeration, which not only increases energy use but also necessitates packaging that further adds to environmental impact. Studies suggest that transportation alone can account for up to 11% of a food item's carbon footprint, making local choices more sustainable (Weber & Matthews, 2008).

In addition to being local, seasonal foods are grown in alignment with natural growing conditions, requiring fewer artificial inputs like heating, lighting, or chemical enhancements. Seasonal produce is often fresher and more nutritious, as it does not need to be picked early or preserved for long periods. By purchasing seasonal fruits and vegetables at local farmers' markets or joining a community-supported agriculture (CSA) program, consumers can reduce their carbon footprint and support local farmers. CSAs, in particular, provide fresh, in-season produce directly from farms to consumers, allowing participants to enjoy high-quality food while fostering local food systems and reducing reliance on imported goods.

Reducing Food Waste

Reducing food waste is another significant way that consumers can lower their environmental impact. In the U.S., approximately 30-40% of food is wasted, with household waste representing a large portion of this loss (Buzby & Hyman, 2012). Food waste has severe environmental consequences: when food decomposes in landfills without oxygen, it produces methane, a GHG with a warming potential 25 times that of carbon dioxide. By managing food consumption and waste more effectively, consumers can play a key role in reducing these emissions and conserving the resources that go into food production.

To minimize food waste, consumers can plan meals in advance, make shopping lists, and purchase only the amounts needed. Proper storage is also essential, as it helps food stay fresh longer and prevents spoilage. Additionally, learning about expiration labels, which are often misunderstood, can help prevent unnecessary waste. Terms like "best by," "sell by," and "use by" are often indicators of quality rather than safety. Understanding that many foods are safe to eat past these dates can help consumers avoid discarding perfectly good food. The USDA's FoodKeeper App is an example of a tool that provides guidance on the storage and safety of various foods, helping consumers make informed decisions about what to keep and what to discard.

Consumers can also consider composting food scraps as an alternative to landfill disposal. Composting converts organic waste into nutrient-rich soil, reducing methane emissions from landfills and creating a valuable resource for gardening. Many communities now offer composting programs, and home compost bins are increasingly popular, allowing individuals to recycle food waste in an environmentally responsible way.

Making Informed Purchases and Supporting Sustainable Brands

In addition to adopting plant-based foods, buying local and seasonal produce, and reducing waste, consumers can further contribute by making informed purchases and supporting brands that prioritize sustainability. Many companies now offer products with certifications that signify environmentally friendly practices, such as organic, Fair Trade, or Rainforest Alliance Certified. By choosing these products, consumers support companies committed to reducing their environmental impact through sustainable sourcing, fair labor practices, and reduced use of harmful chemicals.

Furthermore, consumers can look for products that use minimal or recyclable packaging, helping to reduce waste and pollution. Opting for bulk purchases or items with biodegradable packaging can minimize the use of single-use plastics, which are a significant source of pollution in landfills and oceans. Brands that emphasize transparent sourcing and ethical production provide consumers with choices that align with sustainability values, giving individuals the power to drive demand for greener practices across the food industry.

Advocating for Sustainable Policies

Finally, consumers can play an active role in advocating for policies that promote sustainable food systems. By supporting legislation that encourages sustainable agriculture, reduces food waste, and promotes local food production, consumers can help shape the larger framework within which the food system operates. Engaging in community discussions, voting for sustainability-focused candidates, and participating in local food policy councils are all ways consumers can influence the food system beyond their personal choices. By collectively advocating for sustainable practices, consumers can help shift the food industry toward a more sustainable, equitable, and environmentally responsible future.

In summary, consumer choices—whether in adopting plant-based diets, buying local and seasonal foods, reducing waste, or supporting sustainable brands—have a powerful influence on the food system. When individuals make mindful decisions about what and how they consume, they drive demand for sustainable practices, encourage innovation, and promote environmental stewardship. Through these everyday actions, consumers contribute to a more resilient and sustainable food chain, highlighting the importance of personal responsibility in addressing the global challenges of food production, resource conservation, and climate change.

Best Practices for Producers to Cut Down Emissions and Costs

Producers are in a unique position to reduce the food system's environmental impact by adopting best practices that lower emissions, enhance resource efficiency, and improve long-term sustainability. With access to innovative technologies and farming techniques, producers can make a significant contribution to climate goals while supporting resilient agricultural systems.

Precision Agriculture: A Targeted Approach

One key area for emissions reduction and resource efficiency is precision agriculture, which utilizes data-driven technologies such as GPS, sensors, and satellite imagery to optimize the application of water, fertilizers, and pesticides. Unlike conventional methods, precision agriculture tailors inputs to the specific needs of each part of a field, reducing waste and minimizing the environmental impact of farming. For instance, soil sensors can detect moisture levels, enabling farmers to irrigate only as needed, conserving water. Similarly, GPS-guided systems apply fertilizers and pesticides precisely, reducing excess that can leach into nearby water sources or contribute to nitrous oxide emissions, a potent greenhouse gas.

The environmental and economic benefits of precision agriculture are substantial. Studies show that precision agriculture can reduce pesticide use by up to 30% and water usage by as much as 20% (Zhang et al., 2018). These reductions not only lower input costs for farmers but also decrease the pollution and GHG emissions associated with agricultural inputs. For example, less pesticide use reduces chemical runoff into surrounding ecosystems, protecting biodiversity and water quality. By adopting precision agriculture, producers can enhance yields while also supporting a more sustainable food system that prioritizes resource conservation and minimizes environmental impact.

Soil Health Management: Building Resilient, Carbon-Sequestering Soils

Soil health management is another crucial practice that can lower emissions and improve the resilience of agricultural systems. Healthy soil has a greater capacity to sequester carbon, helping mitigate climate change while enhancing crop productivity and water retention. Practices such as crop rotation, cover cropping, reduced tillage, and the use of organic fertilizers are central to maintaining and improving soil health. Crop rotation, for example, helps break pest and disease cycles, reducing the need for chemical interventions, while cover crops prevent soil erosion and add organic matter, enhancing the soil's ability to retain carbon and water.

The Natural Resources Conservation Service (NRCS) in the United States encourages these practices by providing technical and financial support through programs like the Environmental Quality Incentives Program (EQIP). EQIP helps farmers incorporate soil health practices by offering funding for cover crops, reduced tillage equipment, and organic amendments, supporting long-term soil fertility and resilience (USDA, 2020). By investing in soil health, producers not only reduce emissions but also ensure that their land remains productive and capable of withstanding climate extremes, such as droughts or heavy rainfall, which are becoming more frequent due to climate change.

Renewable Energy Adoption: Powering Farms Sustainably

Producers can also explore renewable energy options to power their operations, which helps reduce reliance on fossil fuels and lowers carbon emissions. Many farms are installing solar panels and wind turbines to provide sustainable power for irrigation, heating, and cooling systems. Solar energy, in particular, is being increasingly used on farms, as solar panels can be installed on rooftops or unused land without impacting crop production. Wind turbines, especially in rural areas with consistent wind patterns, offer another clean energy source that reduces GHG emissions from farm operations. Renewable energy not only reduces carbon emissions but also provides economic benefits by lowering energy costs, enhancing farms' long-term financial sustainability.

Additionally, livestock farms can use anaerobic digesters to capture methane—a potent GHG—from manure and convert it into biogas, a renewable energy source that can power farm operations or even be sold to local power grids. This technology turns waste into a valuable resource, helping farms operate in a more circular and sustainable way. By capturing methane emissions that would otherwise enter the atmosphere, anaerobic digesters can significantly reduce the environmental impact of livestock production. Using renewable energy systems, producers contribute to emissions reductions while reducing operational costs, fostering a more sustainable agricultural sector (Smith et al., 2019).

Water Management and Conservation

Water management is another area where producers can make impactful changes, particularly in regions prone to drought or with limited freshwater resources. Techniques such as drip irrigation and rainwater harvesting allow for more efficient use of water in agriculture. Drip irrigation, for example, delivers water directly to the plant roots, reducing evaporation and waste compared to traditional irrigation methods. Rainwater harvesting systems collect and store rainfall, which can then be used for irrigation during dry periods, reducing reliance on groundwater and surface water resources.

Efficient water management not only conserves a valuable resource but also reduces the energy required for pumping and distributing water, further lowering emissions. By adopting these conservation techniques, producers can reduce water usage by up to 50% in some cases, helping to mitigate the impact of agriculture on local water supplies and ensuring that farming practices are sustainable even in water-scarce regions.

Biodiversity Preservation and Agroforestry

Incorporating biodiversity preservation and agroforestry practices can also benefit the environment and improve farm resilience. Biodiversity on farms, achieved through practices like intercropping, hedgerows, and maintaining natural habitats for pollinators, supports ecosystem services that enhance crop yields, control pests, and improve soil health. For instance, planting diverse crops can attract beneficial insects, reducing the need for chemical pesticides and supporting a balanced ecosystem.

Agroforestry, which integrates trees and shrubs into agricultural landscapes, provides multiple benefits, including carbon sequestration, soil stabilization, and habitat for wildlife. Trees capture CO_2, contributing to climate mitigation efforts, and also provide shade, which can protect crops from heat stress. Moreover, agroforestry practices reduce wind and water erosion, enhancing soil health. By embracing biodiversity and agroforestry, producers create more resilient, productive, and environmentally friendly farming systems.

Producers have an essential role in advancing sustainable agriculture by adopting practices that cut down emissions, conserve resources, and protect biodiversity. Precision agriculture, soil health management, renewable energy adoption, water conservation, and biodiversity-friendly practices are powerful tools for creating a sustainable food system. Through these innovations, producers can increase yields and profitability while contributing to climate change mitigation and environmental stewardship. By leading the way in sustainable farming, producers help ensure that agriculture can meet the food needs of a growing global population without compromising the health of the planet.

Policy Recommendations for a Sustainable and Affordable Food Chain

Policymakers hold significant influence in creating a sustainable food system by implementing policies that support low-carbon practices, enhance resource efficiency, and promote equitable access to healthy food. Through incentives, regulations, and infrastructure investments, policymakers can drive large-scale change across the food supply chain, helping to mitigate climate change while fostering resilient food economies.

Incentives for Sustainable Farming Practices

One of the most effective ways policymakers can support sustainable agriculture is by providing financial incentives that make it easier for producers to adopt eco-friendly practices. Subsidies, grants, and low-interest loans are critical tools for encouraging the implementation of sustainable methods like organic farming, renewable energy adoption, and precision agriculture. By lowering the economic barriers to these practices, policymakers enable farmers to invest in technologies that reduce emissions, conserve water, and improve soil health. For example, the U.S. Environmental Quality Incentives Program (EQIP) offers financial and technical assistance to farmers who incorporate sustainable practices, from cover cropping and conservation tillage to precision irrigation systems. By making these practices financially accessible, EQIP helps reduce the environmental impact of farming and promotes long-term sustainability in the agricultural sector (USDA, 2020).

The economic benefits of such programs extend beyond individual farms; they also strengthen rural economies by reducing input costs, increasing yields, and creating opportunities for new markets in organic and sustainably produced food. Programs that offer incentives for renewable energy use, such as installing solar panels or wind turbines on farms, allow producers to reduce operational costs while lowering emissions. Policies that encourage sustainable farming practices can have a transformative impact, making it easier for producers to adopt low-carbon solutions and helping to build a food system that is both economically and environmentally resilient.

Food Waste Reduction Policies

Food waste reduction policies represent another critical area where policymakers can make an impact. Globally, food waste contributes significantly to greenhouse gas emissions, particularly methane, which is released when food decomposes in landfills. Several countries have pioneered food waste policies that aim to divert edible food from landfills and reduce waste along the supply chain. France, for example, was the first country to pass legislation requiring supermarkets to donate unsold but edible food to charities rather than discarding it. This law has significantly reduced food waste in the retail sector while addressing food insecurity by providing meals to those in need (Mourad, 2016).

Policies encouraging food waste reduction can take many forms, including regulations mandating food recycling and composting, initiatives promoting eco-feed (converting food waste into animal feed), and programs supporting food recovery organizations. Educational campaigns are also essential for shifting consumer behavior and creating a cultural norm around waste reduction. By educating the public about proper food storage, expiration dates, and portion control, policymakers can help reduce household food waste, which is a major contributor to overall food waste. Public awareness campaigns, like those promoted by the USDA, can guide consumers toward waste-reducing behaviors and help create a more sustainable food culture.

Investing in Infrastructure for Local Food Systems

Infrastructure investment is a foundational element for promoting sustainable, locally oriented food systems. Local food systems reduce the distance food travels, or "food miles," thereby cutting transportation-related emissions and supporting regional economies. By building and expanding local food hubs and community-supported agriculture (CSA) programs, governments can support farmers while providing communities with access to fresh, affordable foods. Local food hubs, which aggregate, process, and distribute produce from multiple local farms, can help small-scale farmers reach broader markets and compete with larger, conventional food distributors. This infrastructure also allows institutions such as schools, hospitals, and restaurants to source food locally, promoting a culture of local food consumption (Matson et al., 2013).

In addition to supporting infrastructure, policies that promote collaboration between local farms and community institutions, such as schools and hospitals, can foster food security in underserved areas. Programs that encourage farm-to-institution partnerships help ensure that fresh, nutritious foods are accessible to all residents, including those in food deserts or rural areas with limited access to healthy food. By making local food systems more robust, policymakers can enhance food security, reduce transportation emissions, and strengthen local economies, creating a food system that is both sustainable and equitable.

Policy Recommendations for a Sustainable Food Future

Moving forward, policymakers can expand these efforts by adopting a range of strategies that promote sustainability across the entire food system. Key recommendations include:

1. **Expanding Financial Incentives for Sustainable Agriculture:** Increasing funding for programs like EQIP and similar initiatives in other countries can make sustainable practices more accessible for farmers, especially small-scale producers. Policymakers can also consider tax breaks and direct subsidies for farms that implement regenerative practices, such as soil conservation and agroforestry, which contribute to carbon sequestration and biodiversity preservation.

2. **Establishing National Food Waste Reduction Targets:** Following the examples of France and Japan, which have successfully implemented food waste reduction policies, policymakers can set national targets to minimize food waste across the supply chain. Establishing a food waste hierarchy that prioritizes food recovery, recycling, and composting over landfill disposal can reduce GHG emissions and promote circular economy principles in food systems.

3. **Investing in Research and Development for Sustainable Technologies:** To keep pace with the evolving challenges of climate change, policymakers can increase funding for research into sustainable food production methods, such as lab-grown meat, precision agriculture, and renewable energy solutions for farms. By investing in R&D, governments can foster innovation that will support both current and future food systems.

4. **Supporting Educational Initiatives on Sustainable Consumption:** Public education is essential for driving consumer behavior change. By supporting educational programs

and campaigns, policymakers can encourage consumers to adopt sustainable practices, such as reducing food waste, choosing plant-based foods, and supporting local agriculture. Schools, community organizations, and media outlets can play a role in raising awareness and building a more sustainable food culture.

Consumers, producers, and policymakers each play a vital role in creating a more sustainable and affordable food system. Policymakers, in particular, have the capacity to enact meaningful change by implementing policies that support low-carbon practices, reduce food waste, and strengthen local food systems. By incentivizing sustainable agriculture, promoting waste reduction, and investing in infrastructure, governments can help foster a resilient food future. Through coordinated efforts across all levels of society, we can address the environmental challenges facing the food system, ensure food security, and create a sustainable, equitable, and resilient food system for generations to come.

Chapter 16

A Call to Action for a Balanced, Sustainable Food System

Throughout this book, we have examined the intricate web of economic and environmental impacts associated with the modern food system. Our current food chain—while efficient in many respects—comes with high costs that threaten both economic stability and environmental health. Industrialized agriculture relies heavily on synthetic inputs, energy-intensive production processes, and long-distance transportation, all of which contribute to significant greenhouse gas (GHG) emissions and resource depletion. In 2018 alone, agriculture accounted for roughly 10% of global GHG emissions, primarily from practices that are deeply embedded in conventional food production and distribution (IPCC, 2019).

The economic toll of this system is equally concerning. From the high costs of synthetic fertilizers and pesticides to the substantial waste generated by inefficiencies at every stage of the food chain, we bear the burden of rising food prices, environmental cleanup, and health impacts. Food waste alone costs the global economy hundreds of billions of dollars each year, while also generating methane emissions as food decomposes in landfills (FAO, 2013). These economic and environmental costs underscore the unsustainability of the status quo, highlighting the urgent need for a transition to practices that balance productivity with sustainability.

The Importance of Collaboration Among Consumers, Producers, and Policymakers

Creating a sustainable food system that reduces environmental impact while supporting a strong economy will require a collaborative, multi-stakeholder approach. Consumers, through their choices and purchasing habits, have the power to drive demand for more sustainable products, and their actions, from reducing food waste to opting for plant-based diets, can help reduce the carbon footprint of food. Small but impactful shifts in consumer behavior—like buying local, seasonal produce, and composting organic waste—can make a significant collective difference in reducing emissions and conserving resources.

Producers hold the key to transforming agricultural practices. By adopting technologies like precision agriculture, renewable energy, and soil health management, farmers and producers can reduce input costs, lower emissions, and contribute to the resilience of the food system. Additionally, practices like crop rotation, reduced tillage, and the integration of renewable energy sources, such as solar and wind power, offer pathways to sustainable farming that support both environmental goals and economic viability. While some of these technologies and practices require initial investments, the long-term benefits—including lower operating costs and greater soil fertility—make them worthwhile, especially with the support of public policy and financial incentives.

Policymakers play a pivotal role by setting regulations and offering incentives that support sustainable food practices. Policies such as subsidies for organic farming, grants for renewable energy projects, and food waste reduction programs can ease the transition toward a low-carbon, resilient food system. Internationally, trade agreements that promote environmentally responsible practices and international cooperation on food sustainability initiatives are essential. For example, the Paris Agreement's focus on reducing emissions includes agriculture, recognizing that food systems play a significant role in achieving global climate goals (IPCC, 2019). Policymakers can create the necessary frameworks to empower consumers and producers to contribute to a sustainable food chain, making these changes both feasible and beneficial for all.

Vision for a Future Food System That Is Both Affordable and Low in Carbon Emissions

The vision for a sustainable future food system is one that balances affordability, accessibility, and environmental responsibility. In this future, food systems are designed to produce and distribute food in ways that minimize waste, conserve resources, and emit as little carbon as possible. Affordable access to fresh, nutritious food is central to this vision, and robust local food systems help achieve this by reducing food miles, supporting regional economies, and fostering food security within communities. In this system, food is grown using regenerative practices that restore soil health, increase biodiversity, and sequester carbon, effectively turning agriculture into a solution for climate change rather than a contributor.

Imagine a food system in which urban farms and community-supported agriculture (CSA) programs bring fresh produce to city centers, reducing the need for long-haul transport and providing nutritious options directly to consumers. Precision agriculture and renewable energy power rural farms, making them more productive and less resource-dependent. Innovations like lab-grown meat and plant-based proteins reduce the environmental impact of animal agriculture, while consumers increasingly adopt sustainable eating habits that prioritize plant-based and locally sourced foods. Additionally, food waste is minimized through a combination of consumer awareness, improved distribution networks, and policies that mandate food recovery and composting over disposal.

In this future, policymakers are committed to a circular economy in the food system, where every resource is utilized efficiently, and waste is transformed into new products, energy, or soil-enriching compost. This system supports not only a cleaner environment but also thriving local economies, resilient communities, and enhanced food security.

Sustainable innovations, collaboration across all levels, and a shared commitment to environmental stewardship make this vision possible.

A Final Call to Action

The journey to a balanced, sustainable food system is challenging, but it is essential and achievable with collective action. Consumers, producers, and policymakers must each do their part to address the economic and environmental costs of the current food system and pave the way for a more resilient and equitable future. With commitment and collaboration, we can create a food system that meets the needs of today without compromising the ability of future generations to meet theirs. By making informed choices, supporting sustainable practices, and advocating for policies that prioritize both people and the planet, we hold the power to transform our food system and create a sustainable legacy for generations to come.

Bibliography

Adams, P. (2015). *Food safety and the modern supply chain.* New York: Food World Press.

Al-Kodmany, K. (2018). The vertical farm: A review of developments and implications for the vertical city. *Buildings*, 8(2), 24.

Anderson, L. (2015). *The Green Revolution and its legacy.* Cambridge: Agricultural Press.

Anderson, L. (2018). *Managing water resources in agriculture: Challenges and solutions.* New York: Green Earth Publishing.

Anderson, L. (2019). *Sustainable agriculture: Pathways for resilience in food systems.* New York: Green Earth Publishing.

Anderson, L. (2020). *The Green Revolution revisited: The impact of fertilizers and pesticides on modern agriculture.* New York: Eco-Agriculture Press.

Barlow, C. Y., & Morgan, D. C. (2013). Polymer film packaging for food: An environmental assessment. *Trends in Food Science & Technology*, 31(2), 71-82.

Brown, T. (2019). *Agricultural impacts on soil health and ecosystems.* Boston: Soil Studies Press.

Brown, T. (2017). *Global logistics in food distribution.* London: Supply Chain Insights.

Brown, R., & Harris, M. (2017). *The impact of animal agriculture on biodiversity and ecosystems.* London: Environmental Studies Press.

Brown, T., & Lee, R. (2019). *Chemicals in agriculture: Balancing food security and environmental impact.* Boston: Soil Studies Press.

Brown, T., & Lee, R. (2019). *Nutrient management in modern agriculture: NPK and beyond.* Boston: Agriculture Press.

Buzby, J. C., & Hyman, J. (2012). Total and per capita value of food loss in the United States. *Food Policy*, 37(5), 561-570.

California Department of Food and Agriculture. (2020). Methane reduction policies and incentives for California agriculture. *California Department of Food and Agriculture.*

Carbon Trust. (2019). The carbon footprint of avocados. *Carbon Trust Report.*

Chriki, S., & Hocquette, J. F. (2020). The myth of cultured meat: A review. *Frontiers in Nutrition, 7, 7.*

Coleman-Jensen, A., Rabbitt, M. P., Gregory, C. A., & Singh, A. (2019). Household food security in the United States in 2018. *Economic Research Report No. 270, U.S. Department of Agriculture, Economic Research Service.*

Davis, J. (2016). The economic benefits of home gardening and community-supported agriculture. *Urban Agriculture Journal,* 12(4), 45-56.

Despommier, D. (2010). *The vertical farm: Feeding the world in the 21st century.* Thomas Dunne Books.

Edwards, D., Jones, R., & Brown, S. (2018). Carbon management in commercial food storage and retail. *Sustainable Retail Journal,* 8(3), 102-115.

Edwards-Jones, G., Mila` i Canals, L., Hounsome, N., Truninger, M., Koerber, G., Hounsome, B., Cross, P., York, E. H., Hospido, A., Plassmann, K., Harris, I. M., Edwards, R. T., Day, G. A. S., Tomos, A. D., Cowell, S. J., & Jones, D. L. (2008). Testing the assertion that 'local food is best': The challenges of an evidence-based approach. *Trends in Food Science & Technology,* 19(5), 265-274.

Energy Star. (2020). Energy efficient lighting: How LEDs help reduce energy consumption. *U.S. Environmental Protection Agency.*

EPA. (2020). Food waste basics: How waste management programs help reduce emissions. *U.S. Environmental Protection Agency.*

European Commission. (2018). CAP reform and greening measures for sustainable agriculture. *European Commission.*

European Commission. (2019). EU-Mercosur Trade Agreement: Environmental standards and sustainable agriculture. *European Commission.*

FAO. (2018). *Global Alliance for Climate-Smart Agriculture: Promoting sustainable agricultural practices worldwide.* Food and Agriculture Organization.

Food and Agriculture Organization of the United Nations (FAO). (2013). *Food wastage footprint: Impacts on natural resources.* FAO.

Liptak, A. (2019). The role of municipa

Galt, R. E., O'Sullivan, L., Beckett, J., & Hiner, C. C. (2019). Community supported agriculture is thriving in the United States, but can it grow? *Agriculture and Human Values*, 36(1), 129-140.

Garcia, M. (2017). *Industrial agriculture: Past, present, and future.* Chicago: AgriBooks.

Garcia, M. (2018). *The role of industrial agriculture in global carbon emissions.* Chicago: Carbon Press.

Garcia, M. (2018). *Environmental costs of chemical fertilizers and pesticides in agriculture.* Chicago: AgriEco Publishing.

Garcia, M. (2018). *Chemical use in agriculture: Balancing productivity and sustainability.* San Francisco: Agriculture and Environment Press.

Garcia, M., & Li, P. (2020). *Water footprints and sustainable practices in agriculture.* San Francisco: Water and Agriculture Press.

Garcia, P. (2020). *Livestock emissions and sustainable alternatives.* New York: Green Future Publishing.

Garnett, T. (2011). Where are the best opportunities for reducing greenhouse gas emissions in the food system (including the food chain)? *Food Policy*, 36(Supplement 1), S23-S32.

Green, S. (2016). *From farm to factory: A historical view of food production.* Boston: Heritage Publishing.

Gunders, D., Bloom, J., Berkenkamp, J., Hoover, D., Spacht, A., & Mourad, M. (2017). Wasted: How America is losing up to 40 percent of its food from farm to fork to landfill. *Natural Resources Defense Council.*

Gustavsson, J., Cederberg, C., Sonesson, U., van Otterdijk, R., & Meybeck, A. (2011). *Global food losses and food waste: Extent, causes and prevention.* FAO.

Heller, M. C., & Keoleian, G. A. (2015). Greenhouse gas emission estimates of U.S. dietary choices and food loss. *Journal of Industrial Ecology,* 19(3), 391-401.

Hodge, B. (2017). Waste-to-energy and the reduction of food waste in municipal waste management. *Waste Management Review,* 32(2), 45-53.

IPCC. (2019). Climate change and land: An IPCC special report on climate change, desertification, land degradation, sustainable land management, food security, and greenhouse gas fluxes in terrestrial ecosystems. *Intergovernmental Panel on Climate Change.*

Japan Ministry of Agriculture, Forestry and Fisheries. (2020). The Food Recycling Law and eco-feed initiatives in Japan. *Japan Ministry of Agriculture, Forestry and Fisheries.*

Johnson, A. (2020). *Environmental challenges in animal agriculture.* San Francisco: Animal and Environment Publications.

Johnson, A., & Brooks, E. (2020). *Corporate power in the global food system.* San Francisco: Food Politics Press.

Johnson, A., & Carter, S. (2017). *Organic farming and the path to sustainable agriculture.* San Francisco: Green Earth Publishing.

Johnson, A., & Carter, S. (2017). *Pesticides, fertilizers, and their ecological impact.* Cambridge: Environmental Health Publications.

Johnson, S., Williams, C., & Brown, A. (2015). Cost savings and sustainability of home vegetable gardening: A case study. *Journal of Sustainable Agriculture*, 39(3), 204-219.

Jones, C., et al. (2017). Refrigeration efficiency: New technologies and environmental benefits. *Journal of Sustainable Development*, 10(3), 45-54.

Jones, R., Smith, L., & Thompson, J. (2017). Reducing refrigeration emissions in retail: Innovations and best practices. *Environmental Science and Technology*, 52(6), 2701-2710.

Kamilaris, A., Fonts, A., & Prenafeta-Boldú, F. X. (2019). The rise of blockchain technology in agriculture and food supply chains. *Trends in Food Science & Technology*, 91, 640-652.

King, R. P., Hand, M. S., DiGiacomo, G., Clancy, K., Gómez, M. I., Hardesty, S. D., Lev, L., & McLaughlin, E. W. (2010). Comparing the structure, size, and performance of local and mainstream food supply chains. *Economic Research Report No. 99, U.S. Department of Agriculture, Economic Research Service.*

Krochta, J. M., & Mulder-Johnston, C. (1997). Edible and biodegradable polymer films: Challenges and opportunities. *Food Technology*, 51(2), 61-74.

Lee, R., & Carter, J. (2020). *Regenerative farming and sustainable practices in agriculture.* Philadelphia: Farming Innovations Press.

Lee, R. (2019). *The environmental costs of food production: Water, carbon, and beyond.* Philadelphia: EcoWorld Press.

Lee, R., & Thomas, J. (2019). *Processing the world's food: Advances and impacts.* Philadelphia: Food Engineering Press.

Lee, A., Thompson, J., & Clark, R. (2019). *Agriculture and climate change: Emissions, water use, and land degradation.* Chicago: Climate Impact Press.

Liptak, A. (2019). The role of municipal waste policies in achieving zero waste goals: Case studies from San Francisco. Journal of Environmental Policy, 42(1), 83-95.

Lopez, R. (2019). *The social cost of food production*. Seattle: EarthBound Press.

Matson, J., Sullins, M., & Cook, C. (2013). *The role of food hubs in local food marketing*. U.S. Department of Agriculture.

McKinnon, A. (2018). Decarbonizing logistics: Distributing goods in a low carbon world. *Kogan Page Publishers*.

Miller, D. (2019). *Carbon footprints of crop production*. Washington, DC: Climate and Agriculture Press.

Miller, D. (2020). *Water scarcity and its economic implications in food systems*. Washington, DC: Climate and Agriculture Press.

Miller, D. (2021). *Environmental challenges in modern food production*. Washington, DC: EcoWorld Press.

Minasny, B., et al. (2017). Soil carbon 4 per mille initiative: A global strategy to mitigate climate change and advance food security. *Soil*, 3(4), 237-248.

Mourad, M. (2016). Recycling, recovering and preventing "food waste": Competing solutions for food systems sustainability in the United States and France. *Journal of Cleaner Production*, 126, 461-477.

Mujumdar, A. S., & Law, C. L. (2010). Drying technology: Trends and applications in postharvest processing. *Food and Bioprocess Technology*, 3(6), 843-852.

Poore, J., & Nemecek, T. (2018). Reducing food's environmental impacts through producers and consumers. *Science*, 360(6392), 987-992.

Organic Denmark. (2019). Denmark's Organic Action Plan: Building a sustainable food system through organic farming. *Organic Denmark*.

Roberts, T. (2017). *Agriculture and deforestation: Global perspectives on land use*. London: Forest and Field Publishing.

Roberts, T. (2017). *Deforestation and land-use changes in agriculture*. London: Forest and Field Publishing.

Roberts, T., & Klein, L. (2019). *Food supply in crisis: Global risks and resilience*. Toronto: Crisis and Food Publishing.

Robertson, G. L. (2012). *Food Packaging: Principles and Practice* (3rd ed.). CRC Press.

Smith, J., & Johnson, T. (2018). *Animal agriculture and global emissions: Challenges and solutions*. Boston: AgriScience Publishers.

Smith, J. (2019). *Agricultural water management in drought-prone regions*. Cambridge: University of Agriculture Press.

Smith, J. (2019). *Economic implications of fertilizer and pesticide use in food production*. Washington, DC: Sustainable Agriculture Publications.

Smith, J. (2019). *The environmental consequences of nutrient pollution*. Philadelphia: Water and Land Press.

Smith, J., & Lopez, E. (2021). *Methane emissions and livestock farming*. Cambridge: Livestock and Climate Press.

Smith, J., & Williams, B. (2018). *The interconnected food system*. Oxford: Food Network Publications.

Smith, R., Parker, L., & Nguyen, T. (2019). Renewable energy in agriculture: Assessing the role of wind and solar power. Journal of Sustainable Energy Development, 23(4), 307-321.

Smith, S., Edwards, R., & Brown, J. (2020). Environmental and economic benefits of switching to electric trucks for food distribution in urban areas. *Sustainable Transportation Journal*, 12(3), 401-415.

Smith, T. (2020). Implementing energy management systems in retail: Case studies and results. *Retail Sustainability Review*, 12(1), 35-42.

Smith, T. (2020). Sustainable shopping habits and the future of consumer behavior. *Consumer Sustainability Journal*, 15(2), 40-53.

Specht, K., Siebert, R., Hartmann, I., Freisinger, U., & Sawicka, M. (2019). Urban agriculture of the future: An overview of sustainability aspects of food production in and on urban buildings. *Agriculture and Human Values*, 36(2), 263-285.

Song, J., Murphy, R. J., Narayan, R., & Davies, G. B. H. (2009). Biodegradable and compostable alternatives to conventional plastics. *Philosophical Transactions of the Royal Society B: Biological Sciences*, 364(1526), 2127-2139.

Thompson, J. (2018). *Dietary shifts for sustainability: Trends in plant-based foods and health benefits.* San Francisco: Food and Environment Research Institute.

Thompson, K. (2018). *The industrial age and the transformation of agriculture.* Denver: Industrial Books.

United Nations (2015). Sustainable Development Goals: Goal 12 – Responsible consumption and production. *United Nations.*

United Nations Environment Programme (UNEP). (2020). *Food waste index report 2020.*

USDA. (2020). Environmental Quality Incentives Program: Supporting sustainable agriculture in the U.S. *United States Department of Agriculture.*

van der Zee, B. (2017). The Dutch solution: How the Netherlands became the second-largest food exporter with high-tech farming. *The Guardian.*

Van der Harst, E., & Potting, J. (2013). A critical comparison of ten disposable cup LCAs. *Environmental Impact Assessment Review*, 43, 86-96.

Weber, C. L., & Matthews, H. S. (2008). Food-miles and the relative climate impacts of food choices in the United States. *Environmental Science & Technology*, 42(10), 3508-3513.

Zhang, Q., Wang, C., & Ji, J. (2018). The economic and environmental impacts of precision agriculture. *Sustainable Agriculture Research*, 7(2), 45-52.

Zaman, A., & Lehmann, S. (2013). Challenges and opportunities in transforming a city into a "zero waste city." *Sustainability*, 5(10), 4349-4363.

About the Author

Douglas B. Sims is an environmental professional with over 30 years of experience specializing in NEPA (National Environmental Policy Act) compliance and Environmental Impact Assessments (EIA) and Environmental Impact Statements (EIS) for projects across the United States. His work spans developments from 100-acre infrastructure projects to expansive 500,000-acre land management initiatives, where he has provided critical guidance on environmental impacts, land use, and sustainable development.

Dr. Sims' expertise in regulatory frameworks has been instrumental in advancing projects that balance compliance with environmental standards. His innovative approaches ensure natural ecosystems are protected while meeting developmental goals.

A published researcher, Dr. Sims has contributed to peer-reviewed studies on soil remediation and environmental contamination. His academic knowledge and practical experience make him a sought-after consultant for projects requiring advanced environmental expertise.

Beyond his work, Dr. Sims is passionate about the intersection of environmental science, public policy, and societal challenges. Married to his college sweetheart since the mid-1990s, he and his wife have raised two kids, sharing a commitment to education, environmental responsibility, and lifelong learning.

www.ingramcontent.com/pod-product-compliance
Lightning Source LLC
Chambersburg PA
CBHW071644210326
41597CB00017B/2105